Marine Boilers

Marine Boilers

Third Edition

G. T. H. Flanagan. CEng, FIMarE, MRINA

ELSEVIER
BUTTERWORTH
HEINEMANN

AMSTERDAM • BOSTON • HEIDELBERG • LONDON • NEW YORK • OXFORD
PARIS • SAN DIEGO • SAN FRANCISCO • SINGAPORE • SYDNEY • TOKYO

Elsevier Butterworth-Heinemann
Linacre House, Jordan Hill, Oxford OX2 8DP
200 Wheeler Road, Burlington, MA 01803

First published by Stanford Maritime Ltd 1974
Second edition 1980
Reprinted 1982, 1986
Third edition published by Butterworth-Heinemann 1990
Reprinted 1993, 1995, 1997, 1999, 2000 (twice), 2001, 2003
Transferred to Digital Printing 2004
© G. T. H. Flanagan 1974, 1980, 1990

ISBN 0 7506 1821 3

For more information on all Butterworth-Heinemann
publications please visit our web site www.bh.com

Contents

Preface   vi

1   Stresses in boiler shells   1

2   Auxiliary boilers   7

3   Water tube boilers   27

4   Superheaters and uptake heat exchangers   49

5   Boiler mountings   67

6   Combustion of fuel in boilers   91

7   Boiler operation   105

Index   117

Preface

The aim of this book is to provide practical information on boilers and their associated equipment, as used at sea on steam and motor vessels.

The typical examination questions posed and answered are intended as examples to help the prospective candidate for qualification as Marine Engineer Officer. He should endeavour to produce similar answers based upon equipment aboard his own ship, always stressing safety factors.

In this new edition the opportunity has been taken to include new material on welded boilers and various types of water tube boiler, to deal with the rotary air heater, water level alarm and consolidated type safety valve, and to discuss hydraulic testing and various aspects of survey, maintenance and operational problems.

G.T.H.F.
March 1990

SI UNITS

Mass = kilogramme (kg)
Force = newton (N)
Length = metre (m)
Pressure = newton/sq metre (N/m²)
Temperature = degrees celsius (°C)

CONVERSIONS

1 inch = 25·4 mm = 0·025 m
1 foot = 0·3048 m
1 square foot = 0·093 m²
1 cubic foot = 0·028 m³
1 pound mass (lb) = 0·453 kg
1 UK ton (mass) = 2240 lb = 1016 kg
1 short ton (mass) = 2000 lb = 907 kg
1 tonne (mass) = 1000 kg
1 pound force (lbf) = 4·45 N
1 ton force (tonf) = 9·96 kN
1 kg = 9·81 N
0·001 in = 0·025 mm
(°F − 32) × $\frac{5}{9}$ = °C
1 lbf/in² = 6895 N/m² = 6·895 kN/m²
1 kg/cm² = 1 kp/cm² = 102 kN/m²
1 atmos = 14·7 lbf/in² = 101·35 kN/m²
1 bar = 14·5 lbf/in² = 100 kN/m²

Note: For approximate conversion of pressure units
100 kN/m² = 1 bar = 1 kg/cm² = 1 atmos
1 tonf/in² = 15 440 kN/m² = 15·44 MN/m²
1 HP = 0·746 kW

SI UNITS

Mass = kilogrammes (kg)
Force = newton (N)
Length = metre (m)
Pressure = newton/sq metre (N/m²)
Temperature = degree celsius (°C)

CONVERSIONS

1 inch = 25.4 mm = 0.0254 m
1 foot = 0.304 m
1 square foot = 0.093 m²
1 cubic foot = 0.028 m³
1 pound mass (lb) = 0.4536 kg
1 UK ton (mass) = 2240 lbm = 1016 kg
1 short ton (mass) = 2000 lb = 907 kg
1 tonne (mass) = 1000 kg
1 pound force (lbf) = 4.45 N
1 ton force (tonf) = 9.96 kN
1 kgf = 9.81 N
0.001 in = 0.025 mm
1 °F = (°C × ⁹⁄₅) + 32°
1 btu/hr = 6895 MN/m = 6.895 kN/m
1 kg/cm² = 1 kg/cm² = 102 kN/m²
1 tonne = 1.57 lb/in² = 107.35 kN/m²
1 bar = 14.5 lb/in² = 100 kN/m²

Note: For approximate conversion of pressure units
100 kN/m² = 1 bar = 1 kg/cm² = 1 atmos
1 tonf/in² = 15.440 kN/cm² = 15.44 MN/m²
1 HP = 0.746 kW

CHAPTER 1

Stresses in Boiler Shells

Q. Sketch a double butt strap joint for a multi-tubular tank boiler. State why this must be the strongest joint in the shell.

A. Consider a thin cylindrical shell subjected to an internal pressure. This sets up stresses in the circumferential and longitudinal axes which can be calculated as follows:

CIRCUMFERENTIAL STRESS

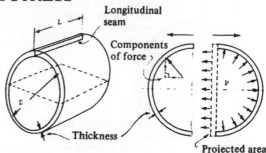

Fig. 1   Stress in the longitudinal seam

If pressure acting upon the circumference is resolved into horizontal components, the resulting horizontal force = pressure × projected area. This is resisted by the stresses set up in the longitudinal joint.

Then for equilibrium conditions:

Horizontal forces to left = Horizontal forces to right

Horizontal forces to left = Resisting force in longitudinal joint

Pressure × projected area = Stress × cross sect. area of joint

Press. × dia. × length = Stress × 2 × thickness × length

$$\frac{\text{Pressure} \times \text{diameter}}{2 \times \text{thickness}} = \text{Stress in longitudinal joint}$$

LONGITUDINAL STRESS

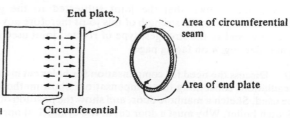

Fig. 2   Stress in the circumferential seam

The force acting upon the end plate is resisted by the stress set up in the circumferential joint.
  Then for equilibrium conditions:

Horizontal forces to left = Horizontal forces to right
Pressure × end plate area = Resisting force in circumferential joint

$$\text{Press.} \times \frac{\pi}{4} \times \text{diameter}^2 = \text{Stress} \times \text{cross sect. area of joint}$$

$$\frac{\text{Pressure} \times \text{diameter}}{4 \times \text{thickness}} = \text{Stress in circumferential joint}$$

Thus it follows that longitudinal joint stress is twice the circumferential joint stress.

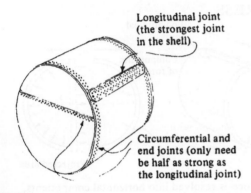

Longitudinal joint
(the strongest joint
in the shell)

Circumferential and
end joints (only need
be half as strong as
the longitudinal joint)

Fig. 3   Riveted joints in a boiler shell

Then rearranging the shell formula in terms of pressure

$$\text{Max. pressure} = \frac{\text{Max. stress} \times 2 \times \text{thickness}}{\text{diameter}}$$

This pressure will be reduced by the efficiency of the joint subjected to the greatest stress.

$$\text{Max. working press.} = \frac{\text{Max. stress} \times 2 \times \text{thickness}}{\text{diameter}} \times \text{Joint efficiency}$$

It has been shown that the joint subjected to the greatest stress will be the longitudinal joint; the strength of this joint therefore governs the allowable working pressure, and so the strongest type of riveted joint used in the boiler is used for this joint. See Fig. 4 on facing page.

Q. Discuss the need for compensation for holes cut in the shell of a boiler. State the regulations concerning this compensation. Show methods of compensation that can be used. Sketch a manhole door, and show the position of these doors in the shell of a Scotch boiler. Why must a door cut in the cylindrical portion of the shell be placed in a certain way?

2

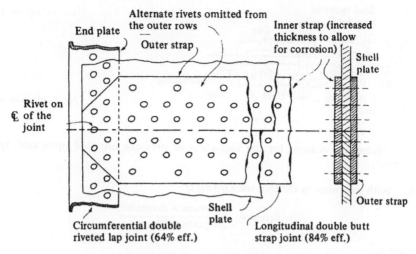

Fig. 4   Detail of riveted longitudinal joint

A. Any material cut from the shell will weaken it by an extent related to the amount cut out. In general any holes cut in the shell, with a diameter greater than:— $2 \cdot 5 \times$ plate thickness + 70 mm must be provided with compensation for the loss of strength due to the material cut out. Some forms of compensating pads are shown in Fig. 5.

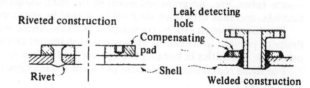

Fig. 5   Compensating pads

The largest holes cut in the shell include the manholes, and where these are cut in the cylindrical portion of the shell they must be arranged with their minor axis parallel to the longitudinal axis of the boiler. This is due to the stress acting upon the longitudinal seam being twice that acting upon the circumferential seam. Thus the shell must not be weakened more than necessary along its longitudinal axis.

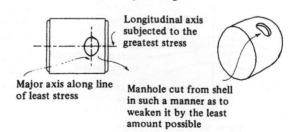

Fig. 6   Arrangement of manhole in a Scotch boiler shell

3

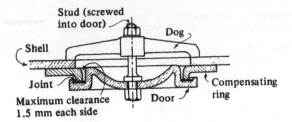

Stud (screwed into door)

Dog

Shell

Joint

Maximum clearance 1.5 mm each side

Compensating ring

Door

Fig. 7   Manhole door

Q. Discuss the reasons for the limitation of pressure imposed upon tank type boilers.

A. With reference to the thin shell formula:

$$\text{Stress} = \frac{\text{Pressure} \times \text{diameter}}{2 \times \text{plate thickness}}$$

Dia

Thickness

Fig. 8

Thus it can be seen that if the stress in the material is to be kept within fixed limits (as is the case with boiler material) then, if the pressure or diameter increases, the plate thickness must also change if the ratio is to remain constant.

Therefore if boiler pressure is increased, either the boiler shell diameter must decrease, or the boiler scantlings increase; the latter leading to increased cost and weight. In order to accommodate the combustion chamber, smoke tubes, etc., no great reduction in the shell diameter of tank type boilers is possible, and thus very thick shell plates would be required for high pressure.

The furnace must also be considered, as its thickness must be kept within certain limits to prevent overheating. However, its diameter cannot be reduced too much, otherwise difficulties in burning the fuel in the furnace would arise.

For these reasons the maximum pressure in tank type boilers is limited to about 1750 kN/m².

Q. Show the reason for the staying of any flat surfaces in a pressure vessel. How can the use of stays be avoided?

A. When a force is applied to a curved plate as shown in Fig. 9, internal forces are set up which enable the plate to withstand the force without undue distortion.

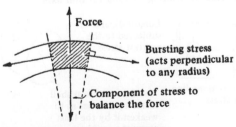

Force

Bursting stress (acts perpendicular to any radius)

Component of stress to balance the force

Fig. 9   Stress in a curved plate

The bursting stress can be resolved into perpendicular components, one of which will oppose the force. The surface will bend until this component balances the pressure. It will then be found that the surface is in the form of an arc of a circle.

When the pressure acts upon a flat plate, it will tend to bend the plate until equilibrium is obtained. Thus to prevent undue distortion the flat

4

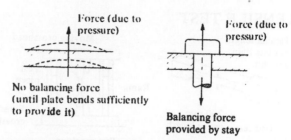

Force (due to pressure)

No balancing force
(until plate bends sufficiently
to provide it)

Force (due to pressure)

Balancing force
provided by stay

Fig. 10

plate must be very thick or supported by some form of stay. It follows that if the use of flat surfaces can be avoided in the design of a pressure vessel there will be no need to fit internal stays. Thus pressure vessels are often given hemispherical ends but, if this is not possible, any flat surfaces must be stayed or of sufficient thickness to resist the pressure without undue distortion.

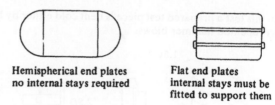

Fig. 11

Hemispherical end plates
no internal stays required

Flat end plates
internal stays must be
fitted to support them

Q. State the regulations relating to the materials used in boiler construction. To what tests must these materials be subjected?

A. The Department of Trade and the classification societies, such as Lloyds, have very strict rules governing the dimensions and materials used in the construction of pressure vessels, so that they can withstand the forces set up by pressure and thermal effects.

The DOT require that carbon steel used in the construction of pressure vessels is to be manufactured by open hearth, electric, or pneumatic processes such as LD, Kaldo, etc. These may be acid or basic in nature. To ensure the materials used are of uniform quality within the requirements laid down, tests are carried out on samples of material.

Some of the materials are as follows: Shell plates for riveted construction, and steam space stays are to have a tensile strength between 430–560 MN/m², with a percentage elongation of not less than 20 per cent on a standard test length. In the case of shell plates for welded construction the strength requirements lie between 400–450 MN/m². Plates which have to be flanged have lower strength requirements, but must have greater elongation; therefore, the tensile strength of these plates should lie between 400–460 MN/m², with a percentage elongation of 23 per cent on a standard length. Similar requirements are laid down for plates, other than shell plates, which are to be welded, also for plates used in the construction of combustion chambers, and the material used for combustion chamber stays.

To maintain control over the final product the following tests are carried out:

TENSILE TEST

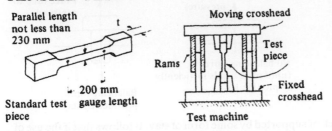

Parallel length
not less than
230 mm

t

200 mm
gauge length

Standard test
piece

Moving crosshead

Rams

Test
piece

Fixed
crosshead

Test machine

Fig. 12 Tensile test

For this standard test pieces are prepared from samples of material; these are then placed in a tensile test machine and loaded to the required values. This enables both the tensile strength and the percentage elongation of the material to be determined.

BEND TEST

In this test a prepared test piece is bent cold either by hydraulic or other pressure, or by repeated hammer blows.

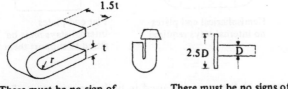

1.5t

t

r

2.5D

D

There must be no sign of
fracture after bending

There must be no signs of
cracking at edges

Fig. 13 Bend and ductility tests

TESTS ON RIVET BARS

Rivet bars, in addition to the tensile and bend tests described, are also subjected to dump testing and sulphur printing. In the latter, tests are carried out on a cross section of the bar to prove there are no sulphur segregates present in the core. In dump testing short lengths, equal to twice the diameter of the bar, are cut from the bar and compressed to half their original length without signs of fracture.

Finally the following tests are carried out on a few completed rivets, selected at random from each batch. The rivet shank must be bent cold until the two parts touch, without any signs of fracture at the outside of the bend. In the other, the rivet head is flattened until its diameter is equal to 2·5 × original diameter, with no sign of cracking at the edges.

CHAPTER 2

Auxiliary Boilers

Q. Sketch and describe a Scotch boiler. Indicate overall dimensions, and plate thickness.

A. For many years the most common boiler in use at sea was the tank type boiler known as the multi-tubular cylindrical, or Scotch boiler. At the present time this type has for the most part been relegated to auxiliary duties.

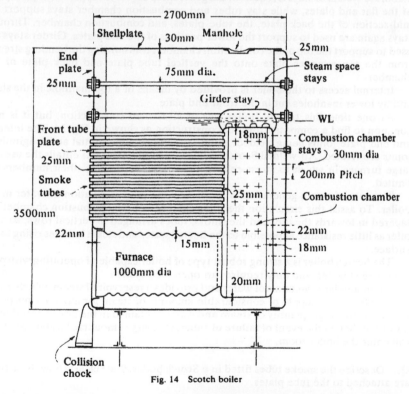

Fig. 14 Scotch boiler

The main components of a Scotch boiler consist of a cylindrical shell containing the furnaces; these are about 1m in diameter, the number depending upon the boiler diameter. Two furnaces are usually fitted for boiler shell diameters up to 4 m, while three furnaces are fitted for diameters greater than this.

The fuel is burnt in these water cooled furnaces, constructed in the form of corrugated cylinders, which due to their relatively large diameter have to be fairly thick and so are liable to overheating problems. Care must be taken to ensure undue deposits of scale are not allowed to build up on the water side of the furnace as this would lead to overheating of the furnace metal.

After leaving the furnace the hot gases enter the combustion chamber; this is also surrounded by water and so again provides for the generation of steam. The top of this chamber is close to the water level in the boiler and if this is allowed to fall below a certain minimum value, overheating may occur; this can lead to distortion and possible failure.

From the combustion chamber the gases pass through smoke tubes, which consist of plain and stay tubes—the latter being necessary to support the flat tube plates. After leaving the tubes, the gases enter the smoke box, and then the uptakes. In many cases gas/air heaters are fitted to increase the boiler efficiency.

As the boiler has a considerable amount of flat surface subjected to pressure, an elaborate system of stays is required. Large steam space stays support the upper parts of the flat end plates, while stay tubes and combustion chamber stays support the mid-section of the back plate, the tube plates, and combustion chamber. Through stays again are used to support the lower portions of the end plates. Girder stays are used to support the top of the combustion chamber, transmitting the bending stresses from the top wrapper plate onto the vertical tube plate and back plate of the chamber.

Internal access to the boiler is provided by means of a top manhole in the shell, and by lower manholes cut in the front end plate.

At one time this type of boiler was of all-riveted construction, but it is now common to find a composite form of construction as shown in Fig. 14. The internal and end plate joints being welded, while the shell circumferential and longitudinal joints are riveted. All-welded Scotch boilers are built but, as this entails the use of a large furnace in which to stress-relieve the completed boilers, their numbers are limited.

The rate of steam generation is limited by the poor circulation of water in the boiler. To assist this circulation the sides and back of the combustion chamber are tapered in towards the top, and the smoke tubes arranged in vertical rows, so as to offer as little resistance as possible to the heated water and steam bubbles rising to the surface.

The Scotch boiler is a strong robust type of boiler capable of operating with poor quality feed water, and only requires an open feed system.

It contains large amounts of water and provides a reservoir of steam, which makes it suitable for the supply of considerable amounts of steam for auxiliary purposes. However, the large quantity of steam and water contained in the boiler introduces a hazard in that, in the event of failure of tubes, etc., large amounts of water and steam can enter the boiler room.

Q. Describe the smoke tubes fitted in a Scotch boiler. State clearly how these tubes are attached to the tube plates.

A. The hot gases leaving the combustion chamber pass through smoke tubes fitted between the back, or combustion chamber, tube plate, and the front tube plate, on the way to the uptakes. Two types of tubes are used: Plain and stay tubes.

The plain, or common tubes are expanded into the tube plates at both ends. The tubes have a diameter of about 65 mm. with a thickness of 5 mm.

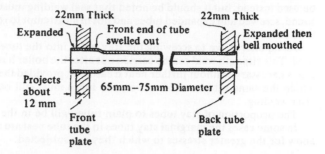

Fig. 15   Plain tube

The front end of the tube is often swelled out to allow for easier tube removal, if this should prove necessary, as even slight deposits of scale forming on the water side of the tube would make the withdrawal of parallel tubes through the holes in the front tube plate difficult. Another provision, sometimes made, is to allow about 12 mm of tube to project from the front tube plate so that if the tube burns away and becomes thin at the back end it can be driven in and re-expanded. However for this to be successful the outside of the tube must be clean and in good condition as corrosion, or scale getting between the faces to be expanded, prevents the formation of a sound, tight seal between tube and plate.

The back end of the tube is bell-mouthed after expansion or, as an alternative, it may be spot welded.

The flat tube plates must be supported, so stay tubes must be fitted, these being screwed and then expanded into both tube plates. The thickness of these stay tubes

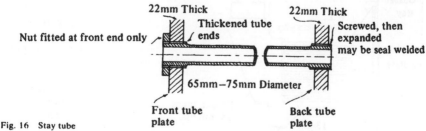

Fig. 16   Stay tube

varies according to the load to be supported, but must not be less than 5 mm as measured at the base of the thread. To prevent undue thickness of metal along the length of the tube, the tube end is thickened by a process known as 'upsetting'. The enlarged ends are then annealed, and finally cut with a continuous thread so that the tube can be screwed simultaneously into both tube plates.

After the tube has been screwed and then expanded into the tube plates, nuts are usually fitted at the front end, but not in the combustion chamber as this would tend to cause overheating due to excess metal thickness. In some cases seal welding may be used instead, but it should be noted that seal welding must only be carried out on sound, screwed and expanded tubes and not in an attempt to repair a corroded or thin tube end.

As an alternative to screwing the stay tubes into the tube plates, welding can be used. This strength weld can be carried out after the boiler has been stress relieved, if this is necessary, without further heat treatment, provided the tubes are not adjacent within the same tube nest. However, the tubes must still be expanded before and after welding.

The proportion of stay tubes to plain tubes will be in the order of 1:3.

In some cases the marginal stay tubes in the tube nest are made slightly thicker to allow for the greater stresses to which they are subjected.

Q. Sketch and describe a vertical smoke tube boiler suitable for auxiliary purposes.

A. The Cochran is a typical tank boiler of vertical type suitable for producing relatively small amounts of low pressure steam for auxiliary purposes.

The fuel is burnt in a furnace having a seamless hemispherical crown, attached to the boiler shell by means of an ogee ring. The products of combustion pass from this

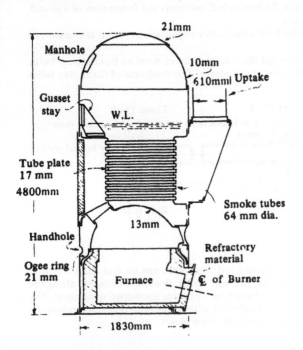

Fig. 17   Cochran smoke tube boiler

furnace into a combustion chamber lined with refractory material, and then through smoke tubes into the smoke box at the front of the boiler.

The cylindrical boiler shell with its hemispherical crown, together with the hemispherical furnace forming the bottom of the pressure space, requires no stays. However the combustion chamber top requires support, and this is provided by means of a gusset stay which transfers the stresses from the flat top of the chamber onto the boiler shell. The flat tube plates are tied together by means of stay tubes screwed into them.

A modified form of this boiler is now produced in an all-welded form.

As all the heating surfaces are below the water level, and the straight, relatively short smoke tubes are readily accessible for cleaning or renewal, this type of boiler forms a robust unit suitable for use with an open feed system and poor quality feed water.

The boiler can be operated with either solid or liquid fuels, although in sea-going vessels oil firing is invariably used.

The Cochran boiler can also be adapted for use as an exhaust gas boiler, using the exhaust gases from an internal combustion engine for the generation of steam. In other cases boilers designed for composite firing, using exhaust gases and/or oil firing, are fitted.

Refractory material is fitted in the combustion chamber, and on the floor and sides of the furnace. Special care should be taken in the latter case to ensure that the refractory is high enough to prevent direct radiant heat coming onto the ogee ring, and lower parts of the hemispherical furnace crown. These lie at the bottom of a narrow water space into which suspended solids from the boiler water tend to settle out, and then form scale on heated surfaces, thus leading to the overheating and subsequent distortion of the ogee ring.

Internal access to the boiler is provided by a manhole in the top of the shell, while handholes in the lower section of the shell provide access to the lower parts of the water space for cleaning and inspection.

Hinged smoke box doors give access to the tubes and tube plate at the front, while a removable rear panel fitted to the combustion chamber gives access to the back tube plate.

Q. Compare a welded type auxiliary boiler with a design using riveted construction.

A. At one time welding was not considered suitable for the manufacture of pressure vessels, but with modern welding and test procedures it has entirely replaced riveting in the design of contemporary boilers.

The use of welded joints avoids many problems associated with riveted joints, such as leakage, caustic embrittlement with resultant cracking between rivet holes, etc. In addition, as there is no longer the need to overlap plates in forming seams, there is an overall reduction in boiler weight; also, problems of overheating where the overlaps occurred on heated surfaces are avoided.

Fig. 17 shows a riveted boiler with a hemispherical furnace attached to the boiler shell by means of the ogee ring connection necessary with riveted construction. This has been replaced in the welded design shown in Fig. 18 by a water cooled furnace of spheroidal shape. This, apart from the burner quarl, needs no protective refractory material, which in turn allows the fully water cooled furnace to provide a greater radiant heat surface with a resultant increase in steam generation. The smaller amount of refractory helps to reduce boiler maintenance, but does give the disadvantage that flame stability is reduced due to the lack of heat reflected back from refractory to stabilise the flame. Thus flame failure devices are essential with fully water cooled furnaces.

The method of attachment by ogee ring causes the bottom of the water space to be very narrow, and so liable to a build-up of deposits, which in turn can result in severe overheating in this area of the boiler; this may be aggravated by direct flame impingement if the protective refractory over the ogee ring has been allowed to deteriorate. The sharp bends of the ogee ring also make it vulnerable to fatigue cracking if the proper steam-raising procedures for the boiler are not followed.

Welding the stay tubes to the tube plates helps to prevent the overheating caused by the thickened tube ends and the use of backing nuts required with screwed stay tubes.

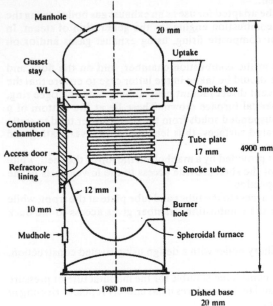

Fig. 18   Cochran spheroid vertical boiler

Q.   Sketch and describe a vertical tank boiler using thimble tubes. Give the plate thicknesses.

A.   This type of boiler was developed by Clarkson, who found it possible to generate steam by causing a prolonged series of spasmodic ebullitions to take place in a series of horizontal tapered thimble tubes heated externally, without any special means being provided for circulation within the tube.

The basic form of the boiler consists of an outer shell enclosing a cylindrical furnace surmounted by the combustion chamber into which the thimble tubes project. These thimble tubes are expanded and bell-mouthed into a cylindrical tube plate forming the combustion chamber. The upper end of this tube plate is connected by a dished plate to the uptake which passes through the steam space, and is then attached to the domed top of the outer shell.

The bottom of the tube plate is attached to the furnace; this in many cases is corrugated, and in turn is connected to the outer shell by means of an ogee ring. Thus the design of the boiler is such that no stays are required. In many cases the joints of these boiler internals are welded.

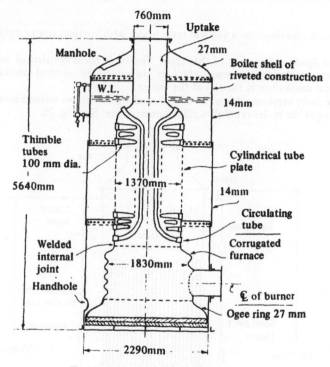

Fig. 19   Clarkson thimble tube boiler

Some versions of these boilers have a removable gas baffle fitted within the combustion chamber to direct the flow of gases from the furnace onto the thimble tubes; in other cases, a number of bent circulating tubes, expanded and bell-mouthed into the tube plate at both ends, are fitted in the gas path as shown in Fig. 19.

The type of boiler shown is oil fired, but other versions can be used as waste heat boilers making use of the hot exhaust gases from internal combustion engines, or designed for composite firing using exhaust gases and/or oil firing as required.

Access to the boiler internals is by means of a manhole at the top of the shell, while cleaning and inspection of the lower parts of the water space is provided by handholes in the shell.

These boilers will operate for long periods without internal cleaning although, if an undue amount of scale forms inside the thimble tubes, it is very difficult to remove. Thus reasonable quality feed water should be provided. The formation of scale will subject the thimble tubes to a certain amount of overheating, but the fact that they are attached at one end only greatly reduces the possibility of tube failure.

The gas side must be kept clean as even thin deposits of soot can cause a drastic reduction in the amount of heat conducted. It is claimed that oily deposits can be burnt off the outside of the thimble tubes when the boiler is dry without damage, but this procedure must not be carried out if circulating tubes attached at both ends are fitted.

Q. Sketch and describe a vertical tank boiler with vertical smoke tubes.

A. The Spanner boiler is of this type. The basic form consists of an outer shell enclosing a cylindrical furnace which is connected by vertical smoke tubes to a cylindrical smokebox at the top of the boiler.

In the early versions of this boiler a riveted shell encloses welded internals, but in later designs the boiler is off all-welded construction. See Fig. 20.

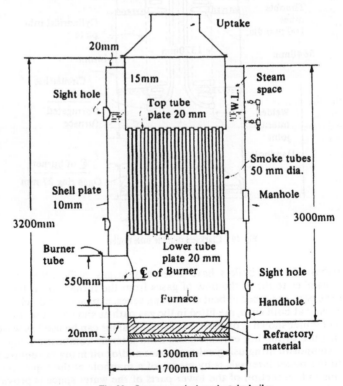

Fig. 20 Spanner vertical smoke tube boiler

The vertical smoke tubes are a patent design known as swirlyflow tubes; they have a special twist along the greater part of their length, only a short portion at each end being left plain to allow for expansion. It is claimed that these tubes are more efficient than normal plain smoke tubes in that they cause the gases passing through to swirl, so coming into more intimate contact with the tube wall and therefore increasing the rate of heat transfer.

No stays are required for the outer shell and for the internals. Only the flat tube plates need to be supported; this is done by stay tubes, of plain section, expanded and then welded into the tube plates.

Internal access is obtained by means of a manhole in the outer shell, and by handholes and sightholes cut at strategic positions in the shell to allow for cleaning and inspection.

The boiler shown in Fig. 19 is oil fired, the fuel being burnt in the water cooled furnace—only the floor of which is covered by refractory material. The gases leaving the furnace pass to the smokebox via the vertical smoke tubes.

In addition to the oil fired version shown, two others are available suitable for use in waste heat recovery systems as fitted in motor vessels. Both of these have the same form of vertical tube stack, which in the exhaust gas version is circulated by the engine exhaust gases, while in the composite version a baffle separates the exhaust gas section of the tube stack from the oil fired flue gas section. These two gas streams are isolated from each other, each leaving by its own uptake, thus allowing oil firing to be used in conjunction with the exhaust gases from the engine.

Q. Sketch and describe a vertical water tube boiler suitable for auxiliary purposes.

A. The basic layout of an Aalborg Q9 auxiliary water tube boiler is shown in Fig. 21. It combines the convenient compact design of a vertical boiler with the greater evaporation rate and flexibility of output associated with water tube boilers.

The fuel is burnt in a furnace surrounded by water walls which, by receiving radiant heat from the furnace flame, increase the evaporation rate of the boiler, while at the same time greatly reducing the amount of refractory material required for the furnace. After leaving the furnace the hot combustion gases flow over the vertical water-filled generating tubes, giving up heat before leaving tangentially to enter the boiler uptake.

Positive circulation of water through heated tubes is provided by a number of external downcomers taking water from the steam drum and feeding it to a circular header at the base of the boiler. This header then distributes the water to the water wall tubes, through which it rises upon heating to enter the annular water space arranged around the flue pipe. From here the water passes up through the numerous generating tubes to enter finally the cylindrical steam drum at the top of the boiler. Refractory material is used to protect the furnace floor and the bottom header from radiant heat. At the top of the furnace the water wall tubes curve over to protect the annular water space from the direct heat of the flame.

Of welded construction, the boiler has a number of flat surfaces which require support. Stays are fitted in the steam drum and in the annular water space, which together with a number of stay tubes provide the necessary support.

Manholes provide internal access to steam and water spaces, with bolted panels giving access to the gas spaces.

A flanged skirt enables the boiler to be secured in place, with additional support from bulkheads etc. to anti-rolling lugs welded to the upper part of the boiler.

15

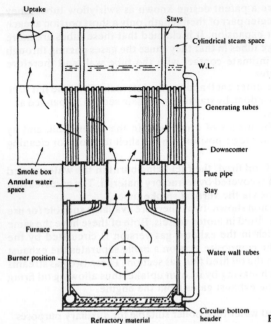

Fig. 21   Aalborg vertical water tube boiler

In many cases the steam drum also acts as a steam receiver for an exhaust gas boiler.

A modified version of this boiler can be used to incinerate waste oil, solid waste and sewage, using the resulting heat to generate steam.

Q.   Sketch and describe a forced circulation boiler suitable for auxiliary purposes. Discuss the use of forced circulation for these duties.

A.   The use of forced circulation results in higher water velocities within the boiler and so permits higher heat exchange rates to be used. Although many forms of forced circulation boilers exist, they all have the basic components indicated in Fig. 22. In the type of boiler shown, water is pumped through the steam generating section where it receives heat from the oil fired furnace. The resulting mixture of steam and water then returns to the steam drum, or receiver, where it passes through a steam separator. The released steam is then used as desired, while the remaining water, together with any incoming feed water is pumped back to the generating section. Two pumps are required, a circulating pump which draws water from the steam receiver and pumps it through the system, and a feed pump supplying water to the receiver. The generating section in this case consists of a single small diameter steel tube arranged in a series of coils, although less extreme types are often used consisting of separate coiled tubes attached to vertical headers, with special proportioning nozzles to ensure that an even supply of water reaches all the coils.

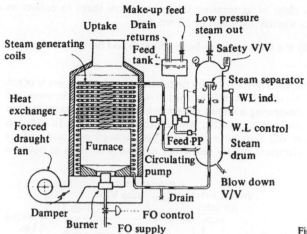

Fig. 22   Forced circulation boiler

Because no fired boiler drum is necessary and only small diameter generating tubes are exposed to the heat of the furnace flame, forced circulation boilers are much safer than tank type boilers.

The type shown contains little more water than that required for the steam load, and so is much smaller and lighter than natural circulation boilers producing comparable amounts of steam, especially where these are of tank type with their large reserves of water. Forced circulation boilers with their small amounts of water and rapid circulation rates permit steam to be raised very rapidly which makes them very suitable for dealing with intermittent steam demands, and because of this they are often used as the steam generating section of package boilers. However these properties also require fairly sophisticated automatic controls to ensure that feed supply etc is matched to the prevailing steam demand. The steam receiver is of all welded construction and has similar mountings to those of a normal boiler drum.

In oil fired versions the high heat exchange rates permissable allows the use of pressurised furnaces if this is desired. The higher combustion pressure leads to higher temperatures so increasing the rate of heat transfer, which would now be too great for a naturally circulated boilder where it would cause overheating.

Waste heat installations, as fitted in motor ships, often use forced circulation exhaust gas heat exchangers in conjunction with orthodox oil fired auxiliary boilers. It should be noted that in this case the water is usually supplied to the bottom of the exhaust gas heat exchanger so giving parallel flow conditions, the water flowing upwards with the gas stream. This although giving less efficient heat exchange than a counterflow arrangement with the water supplied at the top, avoids the problem of vapour locking due to pockets of steam forming in the generator tubes. This could otherwise occur where fluctuating steam demands take place in the presence of a constant flow of exhaust gases from the engine.

Although small diameter tubes and high heat exchange rates are usually associated with the need for good quality feed water, it is claimed that for low and medium pressure ranges the high water velocities due to the forced circulation, flush

17

any precipitated particles clear of generating tubes and allow them to collect as sludge in headers etc from whence they can be blown out.

Q. State what is meant by the term package boiler. Sketch and describe a boiler of this type.

A. Where relatively small, intermittant steam demands are to be met, use is often made of package boilers. This term is usually applied to self contained units mounted on a single bedplate and comprising a steam generating section, feed water system and pump, fuel oil system and pump, together with a forced draught fan. In addition suitable control equipment will also be required. This package now only needs connections to the ship's electrical supply and other necessary services to become operational.

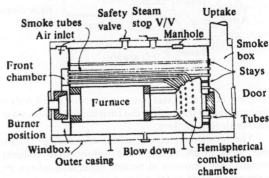

Fig. 23   Typical package boiler

The steam generating section can be of tank or water tube design, using natural or forced circulation. Fig. 23 shows a natural circulation tank type boiler suitable for this duty. The basic arrangement consists of a horizontal cylindrical shell of welded construction, which contains one or two cylindrical water cooled furnaces. These are attached to the front tube plate, and at the rear end to a hemispherical combustion chamber which is pressed out from a single plate. Hot gases from the furnace, having passed through the combustion chamber, then enter the first pass of smoke tubes which conduct them to a front chamber. From here they are directed through a second pass of smoke tubes to a smokebox formed at the rear of the boiler by an extension of the boiler shell beyond the back tube plate. The gases then leave the boiler by means of a suitable uptake.

The general layout is somewhat similar to that of a Scotch boiler and a number of internal stays are required to support the flat tube plates etc. About one third of the smoke tubes consist of stay tubes. Internal access is provided by means of a top manhole in the shell, and by a lower manhole cut in the front end plate. Inspection doors at the rear of the boiler give access to the smokebox and combustion chamber, that of the latter being protected from flame impingement by a removable refractory plug. At the forward end of the boiler is a windbox formed by an extension of the boiler shell beyond the front tube plate. A forced draught fan mounted within this space draws air from the back of the boiler, so that it passes between the outer casing and the boiler shell, thus receiving a certain amount of preheating before it enters the

18

furnace. A control damper is used to regulate this air flow. The usual boiler mountings are fitted.

Automatic controls are provided to regulate the fuel and air supplies in response to changes in steam demand, while at the same time maintaining the boiler water level within the desired limits. Once started up the boiler will continue to operate automatically, flashing up and shutting down to satisfy the steam requirements. The controls are programmed to give a correct sequence for the different operations to be carried out safely. Various safety devices are fitted which will automatically shut the boiler down in the event of loss of water, combustion air pressure, or flame failure. A number of other devices can also be fitted to give warning, and if desired shut the boiler down in the event of high steam pressure or water level, low fuel oil or uptake temperatures etc. When one of these automatic safety cut-outs has operated to shut the boiler down, the device must be manually reset before the boiler can be returned to service after the fault has been rectified.

Although packaging greatly simplifies the specifications required when ordering new tonnage, the advantages to the operator may be limited by the lack of access to be found with some of these package units.

A wide range of package boilers are available, with designs to cover steam pressures from $350 \, kN/m^2$ to $2000 \, kN/m^2$ and with evaporation rates of up to 10000 kg/hr.

Q. Sketch and describe a steam to steam generator. State the purpose of fitting such a unit.

A. These can be used in conjunction with higher pressure water tube boilers to provide low pressure saturated steam for auxiliary purposes. High pressure steam from the water tube boiler is supplied to heating coils placed below the water level in the steam to steam generator and so producing low pressure steam in the generator shell. The steam to steam generator has its own separate feed water system, so protecting the feed system of the high pressure boiler against contamination by aerated and possibly dirty auxiliary drain returns. This is especially important where steam heating is used in cargo tanks or oil fuel heaters, with the ever present risk of the return drains being contaminated with oil from leaking heating coils. If oil should enter the steam to steam generator it would only cause a reduction in the evaporation rate due to fouling of the heating coils, and would not lead to tube failure as could happen if the oil enters a directly fired boiler. However the cost of this arrangement is greater than by supplying the low pressure steam from the high pressure boiler via pressure reducing valves, and if necessary a desuperheater. Thus a steam to steam generator would only be fitted if the risk of oil contamination is considered to outweigh this higher initial cost.

There are two basic designs of steam to steam generators in general use, small vertical units, and larger horizontal ones, a typical layout of which is shown in Fig. 24. The heating coils consist of solid drawn copper, cupro-nickel or mild steel depending upon the operating conditions involved. The tubes are attached to the header by expanding and bell mouthing. The high pressure steam is supplied to the upper part of this header, and then passes through the U-tubes until its latent heat is given up, so producing low pressure steam from the water in the generator shell. This is of welded construction, having a cylindrical form with dished ends so avoiding the need to fit internal stays. Mountings similar to those of a boiler are fitted, the safety valve being

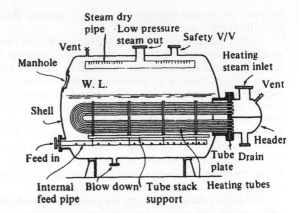

Fig. 24   Steam to steam generator.

able to handle either the normal maximum evaporation, or that produced by the failure of a heating tube. Internal access is provided by a manhole in the top of the shell, and provision is made for the complete removal of the tube stack. The type shown has an evaporation rate of 12000 kg/hr at a pressure of 1000 kN/m² when the heating coils are supplied with 15000 kg/hr of saturated steam at a pressure of 2100 kN/m².

Q.   Sketch and describe a dual pressure boiler capable of supplying low pressure steam for auxiliary purposes.

A.   In some motor ships the auxiliary steam is supplied by means of a dual pressure boiler, a typical arrangement being shown in Fig. 25. This consists basically of an oil fired auxiliary water tube boiler which is used solely to supply steam to the heating elements of a steam to steam generator which then forms the secondary system and produces the low pressure steam required for auxiliary purposes. The primary high pressure system operates on a closed circuit using distilled water, thus giving the advantages that, by reducing the danger of contamination due either to sea water or oil to a minimum, water purity can be maintained without the need for expensive chemical treatment. This allows greater thermal loads to be used in the boiler furnace with little risk of tube failure.

The steam to steam generator which provides the steam for the secondary system, functions in the manner of a high pressure evaporator and, not being directly fired, cannot be damaged by scale deposits, oil contamination or by loss of water. These will only result in a reduction of heat transfer in the generator which in turn leads to an increase of steam pressure in the primary system. Thus the secondary system can operate with poor quality feed water, and can cope with dirty or oil contaminated auxiliary drain returns. However reasonable precautions should be taken to prevent undue amounts of contamination and the secondary feed water system, basically similar to that of a low pressure oil fired boiler, should include a low pressure feed filter and an oil observation tank. The primary system requires no actual feed system, as the condensate from the generator heating coils returns to the steam drum of the oil fired boiler by natural circulation. A small feed pump is required for filling and possible topping up, and this should have a suitable supply of good quality distilled

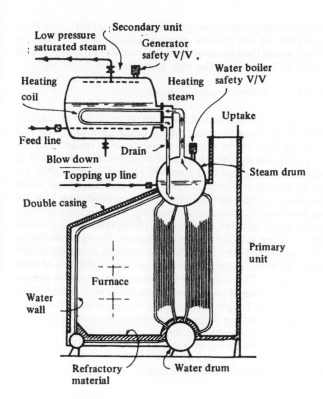

Fig. 25 Dual pressure boiler

water although as the circuit is totally enclosed only a relatively small amount will normally be required to cope with possible loss of water from the primary circuit due to leakage or lifting of the safety valves etc.

In the majority of cases a waste heat system will be involved, the arrangement incorporating an exhaust gas heat exchanger mounted in the main engine uptake, and used to provide low pressure steam when the main engine is operating at or near full power conditions. Again these exhaust gas units operate at relatively low temperatures and so can tolerate the poor quality feed water of the secondary system.

The steam to steam generator is fitted with a safety valve set to lift at the safe working pressure of the secondary system, and able to cope either with the normal maximum evaporation, or that due to a burst heating element, whichever is the greater. The primary system is normally designed to have a safe pressure well above the normal operating pressure. Thus the primary safety valve can be set at a sufficiently high pressure to prevent loss of distilled water in the event of some increase in the primary pressure due to fouling of the heating elements in the steam to steam generator.

Disadvantages of dual pressure boilers lie in increased complexity and higher initial cost, together with greater weight and space occupied than a comparable oil fired auxiliary boiler of orthodox design. Another problem can sometimes arise from overconfidence in the integrity of the closed primary circuit, where cases of contamination have occurred, which then remaining undetected have led to eventual tube failure in the water tube boiler forming the high pressure unit.

Q. Discuss the use of waste heat boilers in motor ships.

A. To increase the overall thermal efficiency of the plant as much use as possible should be made of the heat in the exhaust gases, which in the case of an internal combustion engine will be some 30% of the energy released during the combustion process. In motor ships the fitting of waste heat boilers enables some 20% to 60% of this heat in the main engine exhaust gases to be recovered and used to generate steam. The actual percentage recovered depends upon the complexity of the waste heat system, and obviously some use must be found for the steam produced. This can range from simple heating processes such as fuel oil heating, or domestic hot water and other services, to its use in more complex steam driven plant. Here again is a wide scope, ranging from pumps to steam turbine driven alternators. However the more complex this steam plant the greater will be its initial cost, and the more sophisticated systems will normally only be justified in ships spending at least two thirds of their time at sea at design service power. The output of the waste heat recovery system should be sufficient to handle the normal steam requirements at sea. If additional steam from oil fired boilers would be frequently required to supplement the steam produced by the waste heat boiler, then the fitting of a complex waste heat recovery system to produce steam will not normally be justified.

Another advantage gained from the fitting of waste heat boilers is that they will act as silencers and spark arresters.

Waste heat boilers can be grouped into two main types:

COMPOSITE BOILERS

These consist of auxiliary boilers so arranged that they can receive the heat required for the generation of steam either from the main engine exhaust gases, or by the burning of oil in the boiler furnace. In most cases the oil firing can be used in conjunction with the exhaust gases to support or even replace them. A suitable by-pass line with a change-over valve will be fitted so the main engine exhaust gases can be diverted from the boiler when required.

EXHAUST GAS BOILERS

These consist of some form of exhaust gas heat exchanger mounted in the main engine uptake. However the output of these units is directly dependant upon the main engine output and so it is necessary to provide an additional oil fired boiler to supplement or surplant the steam produced by the exhaust gas unit when the main engine is operating at low load conditions, or when it is stopped. In many cases a convenient arrangement is to use the drum of the oil fired boiler as a steam receiver for the exhaust gas heat exchanger. This gives the advantages that only a single steam drum with its associated mountings is required, and that the oil fired boiler is kept in stand-by condition ready for immediate oil firing to support or replace the heat from the main engine exhaust gases.

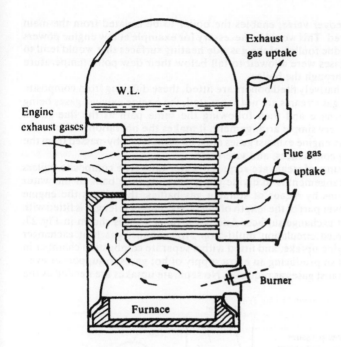

Fig. 26 Composite boiler

Q. Sketch and describe a Composite boiler suitable for producing low pressure steam for auxiliary purposes.

A. The majority of motor ships are fitted with some form of waste heat boiler. The output from these is dependant upon the temperature and mass of exhaust gas available from the main engine, and as it would obviously be unsatisfactory to have steam available only when the main engine was in operation, it is necessary to have an alternative method of steam generation. A common means of doing this is to fit a Composite boiler, this being so arranged that it can be fired either by main engine exhaust gases, or by burning oil in the boiler furnace. In most cases the gas flows are kept separate, each having its own uptake, this permits the oil firing to be used in conjunction with the engine exhaust gases, to support them at low engine loads, or to replace them when the engine is stopped. By this means the output of steam can be maintained independent of the engine power.

Most of the standard tank type auxiliary boilers can be modified for composite firing, a typical example consisting of a vertical, single pass smoke tube composite boiler being shown in Fig. 26. This forms a simple compact unit, there being no need for forced circulating pumps etc. The modification consists of an additional tube nest added to the basic oil fired design, through which engine exhaust gases can be circulated so providing heat for the generation of steam. As the gas flows are kept separate, oil firing can be used to support the main engine exhaust gases at low load conditions. In some versions a return pass of tubes are fitted so that the engine exhaust gases can make two passes through the boiler. A by-pass line is fitted which, with a suitable changeover valve, enables the boiler to be isolated from the main engine exhaust if desired. This would be necessary for example at low engine powers in order to prevent undue fouling of the gas side heating surfaces that would tend to occur if the exhaust gases were allowed to fall below their dew point temperature during their passage through the boiler.

In a few cases alternatively fired boilers are fitted, these differing from composite boilers in that the two gas streams are not separated, the engine exhaust gases being led into the boiler furnace and then following the same path as the flue gases. Although this gives a very simple arrangement, it makes the operation of the boiler much less flexible as the engine exhaust gases must be completely by-passed from the boiler before oil firing commences, and vice versa.

Where larger amounts of steam are required water tube type composite boilers may be used. One arrangement is to divide the generating tube bank of the water tube boiler into sections by means of suitable gas baffles, this allows the engine exhaust gases to flow over part of the length of the tubes. This part length is fitted with fins to extend the heat exchange surface. Another arrangement is shown in Fig. 27. This consists of a forced circulation multi-loop type exhaust gas heat exchanger placed in the main engine uptake, and fitted with a separate combustion chamber in which oil can be burnt so producing an extra supply of hot gases to support or even replace the engine exhaust gases as required. No separate uptakes are needed as the

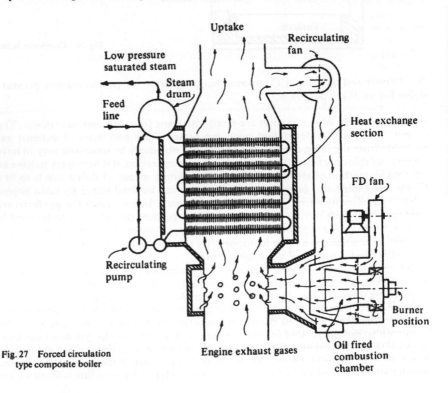

Fig. 27   Forced circulation
type composite boiler

combustion chamber can be pressurised by its own air supply so preventing the engine exhaust gases from interfering with the combustion of the oil.

Composite boilers form a simple waste heat system for the continuous generation of steam for auxiliary purposes, with the advantage of the boiler being kept in operation both at sea and in port, so avoiding long periods of shut down which can often result in corrosion problems.

Q. Sketch and describe a waste heat recovery system including an exhaust gas boiler, capable of supplying the auxiliary steam requirements of a motor ship.

A. Where relatively large amounts of steam are required from a waste heat recovery system it is usual to fit an exhaust gas boiler in the main engine uptake. One or more oil fired boilers will also be necessary to supply steam as required when the engine is running at low power or stopped. The various steam raising units can be of tank or water tube design, using natural or forced circulation. The latter provides many advantages in waste heat installations, one of these being that the various units can be positioned to best suit the prevailing engineroom layout without having to make provision for natural circulation.

Many arrangements are used, the most simple consisting of separate exhaust gas and oil fired boilers, each having its own feed connection, but each discharging into a common auxiliary steam main. This type of installation often makes use of tank type boilers, and are suitable for providing fairly small amounts of low pressure saturated steam. However this arrangement has some disadvantages, one of which is the problem of thermal stressing that can arise if the relatively large amount of cold water in the exhaust gas unit is not preheated before being put into operation.

Another convenient arrangement in common use is to allow the steam drum of the oil fired boiler to act as the steam receiver for a forced circulation exhaust gas heat exchanger as shown in Fig. 28. This gives the advantages that only a single steam drum is required and that the oil fired boiler is kept in stand-by condition ready for immediate use.

The exhaust gas heat exchanger is a forced circulation multi-loop type, of all-welded construction. Extended heat exchange surfaces are used, similar to those in economisers, mild steel fins being fitted in the higher temperature regions, while cast iron gills are fitted to give protection against corrosion in those sections where the gas stream may fall below its dew point temperature under normal operating conditions. The tube elements may be arranged to provide a staggered gas path in order to obtain better rates of heat transfer, or a straight through path which is less liable to suffer from gas side fouling.

Some of these waste heat recovery systems provide superheated steam for use in turbo-alternators which are capable of supplying the ship's electrical power requirements while at sea and proceeding at or near her normal service speed. In other cases, where only low pressure saturated steam is required, the oil fired unit may take the form of a dual pressure boiler of the type shown in Fig. 25.

In all waste heat steam generating installations some form of control over the amount of steam produced by the exhaust gas boiler is necessary. Various methods are available such as a by-pass arrangement to regulate the amount of engine exhaust flowing through the exhaust gas boiler, or by varying the amount of heat exchange surface in use. Another method is to dump any excess steam produced into the auxiliary condenser, while in other cases the system is designed with a safe pressure

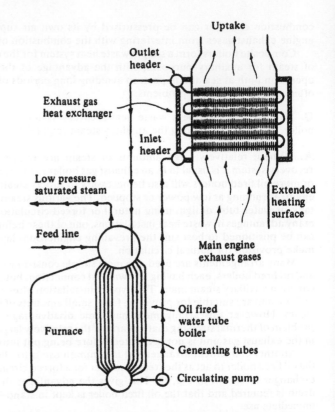

Fig. 28 Exhaust gas unit

well above the normal operating pressure. Then if steam demand is too low to absorb all the heat received by the exhaust gas boiler, its operating pressure will rise so increasing the saturation temperature of the steam and, by reducing the temperature difference between the steam and gas streams, serving to reduce the amount of heat transferred.

CHAPTER 3

Water Tube Boilers

Q. Discuss the reasons for the general adoption of water tube boilers in place of the Scotch boiler for the supply of main engine steam.

A. With the demand for higher efficiencies, steam temperatures have steadily increased, and this for various reasons is coupled with an increase in boiler pressure.

A boiler forms a more efficient heat exchanger if it consists of a large number of small diameter tubes, rather than a small number of large diameter tubes.

A basic design factor involved in these points is the tube diameter; the reasons for this are as follows:

Thin Shell Formula

$$\text{Stress} = \frac{\text{pressure} \times \text{diameter}}{2 \times \text{thickness}}$$

Thus for a given maximum stress, as pressure increases, the diameter must decrease in order to keep the thickness within reasonable limits.

Conduction Formula

$$\text{Heat conducted} = \frac{\text{temperature difference} \times \text{area} \times \text{time}}{\text{thickness}}$$

Other factors remaining constant: as thickness decreases so heat conducted increases.

Surface Area

A number of small diameter tubes offer a greater heat exchange surface per unit length than a comparable large diameter tube, i.e. four 50 mm diameter tubes have approximately equal cross-sectional area to one 100 mm diameter tube, but have twice as much surface area per unit length.

Thus the use of small diameter tube enables higher pressures to be used, while still allowing thin tube walls, which together with the greater heat exchange surface available, enable the heat evolved by the burning fuel to be more readily transmitted to the boiler water—thus allowing high evaporation rates.

In addition the thin walled tubes are easier to manufacture, to bend, and to expand and bell-mouth into drums and headers. They also reduce the weight of metal in the boiler.

27

The small amount of water in the boiler also reduces the overall weight of the boiler and, together with the small diameter tubes, reduces the danger in the event of a tube failure, but gives the disadvantage of little reserve water and steam in the boiler, and thus efficient control is required for water level, etc.

The high rate of heat transfer across the thin tube walls, together with the small bore, demands rapid and positive water circulation through tubes, etc. Water tube boilers can be designed to give sufficient natural circulation up to very high pressures by sloping the tubes by at least 15° from the horizontal, and by the use of unheated downcomers to supply water to the lower parts of the boiler.

The small bore tubes demand pure feed conditions to prevent scale formation, which could lead to blockage and overheating. In addition the thin metal thickness gives little allowance for corrosion, and care must be taken to reduce this to a minimum on both gas and water sides of the boiler. These factors give another disadvantage in that elaborate closed feed systems must be used in conjunction with water tube boilers.

The flexibility of water tube boilers, coupled with their positive circulation, and reduced amounts of refractory material allow for rapid steam raising.

The layout of the water tube boiler permits the furnace to be designed to give efficient combustion conditions, and also allows the boiler shape to be modified to some extent to fit the space available in the ship.

Q.   Describe the construction of the drums for a water tube boiler.

A.   In early water tube boilers the drums were riveted or forged from a single ingot, but for modern boilers they are invariably of welded construction.

The materials and construction procedures are governed by very strict rules laid down by the DOT and the classification societies so that the completed drums can properly withstand the forces set up by internal pressure and thermal effects.

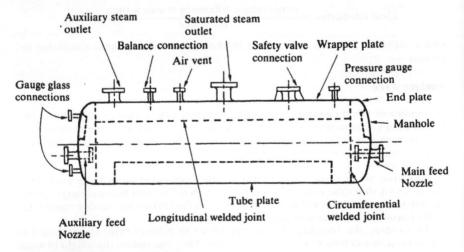

Fig 29.   Steam drum for a water tube boiler

Each drum will consist of four main parts; the cylindrical portion consists of two separate plate sections, called respectively the tube and wrapper plates, and two end plates. These are dished to avoid the use of internal stays.

The thickness of these plates is related in a general way to pressure and diameter by the thin shell formulae, allowance being made where necessary for any reduction in strength due to holes cut in the shell. In large drums the tube plate is made thicker to compensate for the numerous holes drilled in it, the plate edges being tapered down to the same thickness as the wrapper and end plates before welding is carried out.

Before the plates are cut to the required sizes, test pieces are taken and checked to see that the material is satisfactory.

The plates are then cut to size, and bent to the correct curvature. Plates of moderate thickness can be bent cold but, in the majority of cases, the plates forming the drums of water tube boilers are too thick for this and need to be bent hot. It is not possible to bend the plates right to edges forming the longitudinal seams, so the plates are cut wider than necessary, and then the correct width is cut from the curved portion after bending.

After completion of hot bending the plates are allowed to cool. Any mill scale is then removed by sand blasting, or by pickling in acid.

The plate edges are then machined to the correct shape for welding; the tube and wrapper plates are then assembled, being adjusted to their correct positions by means of jacks, and held firmly and accurately in place by clamps which are tack welded in place.

Test pieces, cut from the original plate material, are then lightly welded to each end of the longitudinal seams in such a manner that the weld can be continued across the test piece during the welding of the main seams.

Some form of automatic fusion welding process is then used, the machine, placed vertically above the joint, moving along the length of the seam while the drum remains stationary. When the weld is completed, the drum is rotated until the other joint comes directly under the machine, and the welding process is repeated. Upon completion of the welds, the test pieces are cut from the ends of the joints.

The end plates are now pressed to the required shape and size, the manholes cut and, after cooling, the necessary machining to plate edges, and manhole landing are carried out. The end plates are then placed firmly and accurately in position, and welded to the cylindrical portion of the drum. In this case the welding machine remains stationary, while the drum rotates beneath it, so allowing the machine to weld vertically downwards around the complete circumference.

Upon completion, the seams are dressed flush and non-destructive tests, such as X-ray examination, carried out along the complete lengths of the welds. If any defects, such as porosity, lack of fusion, cracks, or slag inclusions are revealed, the affected sections must be cut out by pneumatic tools and then re-welded. The repaired sections should be again tested and proved sound.

The drum is now marked off; the holes cut for the drum mountings, the branch pieces, etc. being welded in place as required. Holes for the tubes are now made, special drilling machines being used so that the holes are drilled at their correct angle to the plate surface. Any stub tubes to be fitted are then welded in place.

When all machining and welding has been completed, the drum is stress relieved; this is done by heating it to between 600–650°C. It should be maintained at this temperature for a period of time, and then allowed to cool slowly in still air. The test pieces, previously cut from the ends of the longitudinal joints, undergo the same heat treatment. They are then cut up to provide the required test specimens. When satisfactory reports on the results of these tests have been received, the drums are prepared for hydraulic testing. Where no stub tubes are to be welded, the tube holes may be drilled after the hydraulic test has been completed.

Q. Discuss the functions of drums and headers as used in water tube boilers.

A. The vast majority of water tube boilers consist of a steam drum, one or more water drums, and a number of headers. These components are interconnected by numerous tubes.

These drums and headers may be considered as follows:

Steam Drum

In natural circulation boilers the motive power to provide a strong positive circulation is obtained by the difference in density between water at different temperatures.

Thus the steam drum provides a reservoir of relatively cool water giving the gravitational head necessary to displace the high temperature mixture of steam and water, with its much smaller density, from the heated tube surfaces. The drum then provides the space for the separation of this steam and water mixture as it returns. Again the difference in density allows the dry steam to rise, and leave from the top of the drum, while the water joins the incoming feed, to enter the downcomers and so rejoin the circulation circuit within the boiler.

The steam drum thus contains a reserve of steam for manoeuvring purposes. It also receives the incoming feed water so giving the head of water necessary for the proper operation of the boiler, and provides for the distribution of this water to the downcomers.

Water Drum

This provides for the distribution of water, entering it from the downcomers, to the screen and generating tubes and in some cases to the water wall headers.

It also provides a space for the accumulation of suspended solids which may be precipitated from the boiler water. The blow down connection enables these to be blown out as required.

Headers

These perform a similar duty to that of the drums, only size forming a distinction between them. In general the drums are large enough to be entered through manholes, whereas access to the interior of the headers is only provided by handholes.

General Design Features The drum thickness is related to working pressure and diameter in a general way by the thin shell formulae. Thus the smaller the drum diameter for a given pressure, the thinner the plate sections that can be used, giving a lighter and cheaper drum. However, if it is made too small difficulties arise with the separation of the steam and water within the drum. In some cases special steam and water separators have to be fitted to overcome this problem.

Too small a drum can also cause problems with the control of the boiler water level. In some cases this can lead to severe fluctuations of the water level during manoeuvring conditions.

The drums and headers also provide the tube plates necessary for the attachment of the tubes. The drums are normally formed of two complete plate sections forming the cylindrical portion, and in order to provide allowance for the reduction in strength due to the holes drilled in the tube plate, it is usually of greater thickness, being tapered down in way of the welded joints.

The headers may be cylindrical or rectangular in section, the latter being possible due to the small overall dimensions which enable the flat surfaces involved to withstand the stresses imposed upon them by the internal pressure, without having to be of undue thickness.

Q. Give the various types of tubes used in water tube boilers, together with their main functions.

A. The following types of tubes are used in water tube boilers:

Generating Tubes

These consist of numerous small diameter tubes placed in the main flow of hot gases, so forming a large heat exchange surface; the generation of steam takes place mainly by convection.

For a given rate of water circulation the minimum allowable tube diameter is limited, as below a certain value the ratio of steam to water becomes excessive and leads to possible overheating.

Another limitation is imposed by the fact that, while sufficient heat exchange surface must be provided for the gas exit temperature to be low enough to ensure economic operation, if there is too much the gases will fall below their dew point temperature. This will lead to corrosion of the heating surfaces.

In general the number of generating tubes tends to be reduced in modern boilers, until indeed in radiant heat boilers no generating tubes as such are fitted. Water walls receiving radiant heat are used instead.

Screen Tubes

These are placed adjacent to the furnace, so receiving heat from the flame together with heat from the hot gases leaving the furnace; therefore they need a relatively large diameter to keep the ratio of steam to water low enough to prevent overheating.

The duty of the screen tubes is to protect the superheater tubes from the direct radiant heat of the furnace flame.

Water Wall Tubes

These are used basically to contain the heat of the furnace, thus reducing the amount of refractory material required.

In some types of boilers, water cooled refractory walls are used. These consist of tubes with studs welded onto them, covered with refractory material, which can now

31

withstand the high temperatures without damage. In other designs part of the tube surface is exposed to radiant heat which helps to generate steam. In some radiant heat boilers the tubes are welded together along their length by fins or strips, and no refractory is required.

Downcomer Tubes

These consist of large diameter, unheated tubes placed outside the gas stream which act as feeders to the water drum and headers.

Riser or Return Tubes

These return steam and water from the top water wall headers to the steam drum.

Superheater Tubes

These consist of small diameter tubes placed in the main gas stream, after the screen tubes. Their duty is to superheat the saturated steam leaving the drum to a temperature suitable for use in the main turbines. They must be protected from direct radiant heat as they are liable to overheating due to the much smaller specific heat of steam compared to that of water.

Superheater Support Tubes

These relatively large diameter tubes act basically as water cooled supports for the superheater tubes.

The metal surface temperature of all these boiler tubes must be considered and, for all tubes containing water, the working temperature is assumed to be the saturation temperature corresponding to the boiler pressure, plus 15°C. Thus solid-drawn mild steel tubes can be used. In the case of superheater tubes, however, the temperature is considered to be the maximum superheat temperature, plus a value in the order of 30°C for convection type superheaters, and in excess of this figure for radiant heat type superheaters. For steam temperatures above 455°C heat resisting alloy steel, containing small amounts of chrome and molybdenum, must be used.

In the majority of cases boiler tubes are expanded, and then bell-mouthed into drums or headers. For large diameter tubes, such as downcomers, grooved seats are used to assist the expansion in forming a tight seal. Where high temperatures are involved, as with superheater tubes, welding may be used in place of expanding.

There are three main methods of arranging boiler tubes:

Straight tubes are easy to clean and replace but can only be used in conjunction with headers as, if used with drums, the tubes will not enter perpendicular to the tube plate, and the holes would have to be recessed, or arbored, in order to keep the bell-mouthing near the drum surface. These recesses would act as stress risers, and are banned.

Curved tubes can be used with drums, but even here the tubes will not normally enter perpendicular to the tube plate, but at an angle; this makes expanding more difficult, and also results in a thicker tube plate.

In the third method, often referred to as **bent tube**, all the tubes are bent so as to enter perpendicular to the surface of the tube plate. This gives the advantage of easier expansion and bell-mouthing, and also enables a thinner tube plate to be used. The disadvantage lies in the relatively sharp bends, which make cleaning and tube replacement more difficult, and also means more spares must be carried as tube curvature varies from row to row. However, the high quality feed available for modern boilers, which gives little risk of scale formation, enables this method to be used to advantage.

Q. Sketch and describe a header type water tube boiler. Give tube sizes and show the position of the superheater. Indicate the gas and water flows, and give the gas temperatures from furnace to funnel.

A. The header type boiler is robust and suitable for use with poor quality feed; the straight relatively short tubes allow good internal cleaning, and easy tube replacement.

The main disadvantages are the large number of handhole doors in the headers, and extensive furnace brickwork.

Steam generation is mainly by convection, only the lower rows of screen tubes being subjected to radiant heat.

THREE PASS DESIGN

The basic form consists of a single steam drum connected to a series of sinuous headers inclined at 15° to the vertical. Straight tubes, inclined at 15° to the horizontal, expanded and bell-mouthed into these headers, allow the water to circulate to headers at the rear of the boiler similar to those at the front. From here the water now containing a large number of steam bubbles passes back to the steam drum via the horizontal return tubes.

By the use of baffles, the combustion gases are directed so as to make three passes over the generating tubes, the heat from these gases being used to change the water in the tubes to steam.

As the water is heated, so forming a mixture of hot water and steam bubbles, its density decreases, providing the strong circulation through the inclined generating tubes necessary to prevent them overheating.

The required superheat temperatures are obtained by positioning the superheater with more or less tubes beneath it. Indeed for low superheat conditions it is placed above the return tubes at the top of the first gas pass.

A horizontal header, placed across the front of the boiler below the front upright headers and connected to them by short nipples, acts as a mud drum, effectively trapping a large proportion of any sludge in the boiler water. This sludge can then be removed by means of blow down valves fitted at each end of the horizontal header.

The bottom row of inclined tubes consists of 100 mm diameter screen tubes, above these are numerous rows of 46 mm diameter generating tubes. The top horizontal return tubes consist of two rows of 100 mm diameter tubes.

The oil fuel burners are arranged along the front of the boiler, and combustion air passing through the double casing of the boiler enters through air registers into a refractory lined furnace where combustion takes place.

SINGLE PASS DESIGN

This is of the same basic design as the three pass boiler, but with the baffles omitted so that the gases make only one pass over the inclined generating tubes. In general this type of boiler is used for higher steaming conditions than the three pass and, in order to obtain a greater heat exchange surface, smaller diameter tubes are used. The screen tubes now consist of two rows of 50 mm diameter tubes, with rows of 32 mm diameter generating tubes above. The return tubes remain unchanged. As before, access to these tubes is obtained through handholes cut in the headers.

The superheater in the single pass design is always placed low down in order to give the required superheat temperatures.

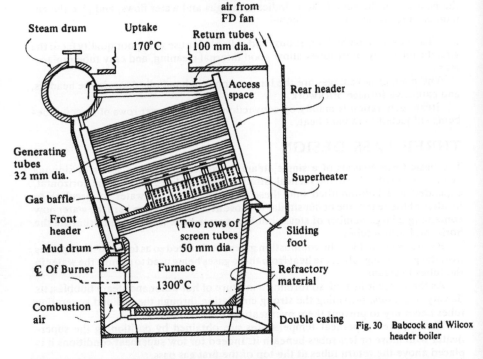

Fig. 30 Babcock and Wilcox header boiler

Water Walls In some cases the more highly rated single pass boilers have water walls fitted at the sides of the furnace. These walls consist of 83 mm diameter, studded tubes, running parallel with the generating tubes between two vertical headers; these in turn are connected to the steam drum by a number of external downcomers and risers consisting of 112 mm diameter tubes.

In addition the largest boilers of this type may have a water cooled rear wall.

In all but the oldest versions, the steam drum is of all-welded construction; the headers are solid forged. All the tubes, and connecting nipples, are solid-drawn mild steel.

The boiler double casing is of bolted construction.

Q. Sketch and describe a two drum, bent tube boiler. Give tube sizes and show the position of the superheater. Indicate the gas and water flows, and give gas temperatures from furnace to funnel.

A. This type of boiler with its largely water cooled furnace is suitable for high operating conditions. The long bent tubes however, due to difficulty of cleaning and tube replacement, demand good quality feed water in order to avoid the formation of scale. The basic design consists of two drums; the larger steam drum is placed above a Smaller water drum and the two are connected by numerous 32 mm diameter generating tubes. See Fig. 31.

The furnace is placed to one side, the hot gases passing over two rows of 50 mm diameter screen tubes to a superheater placed between the drums in front of the generating tube bank.

After leaving these tubes, the gases enter the economiser section in the lower part of the uptakes. Baffles are placed to direct this gas flow as required.

Water walls consisting of 50 mm diameter tubes are used on the roof, side, and rear of the furnace. These water wall tubes are connected to drums or headers as required.

In the Foster Wheeler D-type boiler, 82 mm diameter underfloor tubes are used to supply the lower headers for the water walls with water from the water drum. External downcomers are only fitted between the steam and water drums. Feed water entering the steam drum flows through the downcomers to the water drum to replace the water rising up through the water wall and generating tubes, as its density decreases upon receiving heat from the hot gases, thus providing the positive

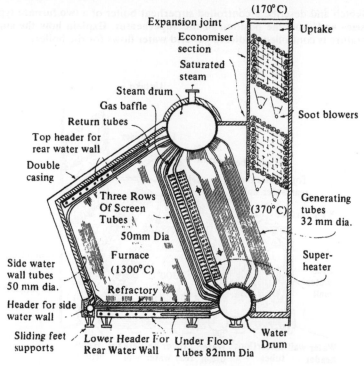

Fig. 31 Foster Wheeler D-type boiler

circulation of water required in all types of water tube boilers. From the top headers, risers return the mixture to the steam drum.

In the Babcock and Wilcox version, The Integral furnace boiler, external downcomers are used not only to supply the water drum, but also the water wall headers; the underfloor tubes in this case are omitted.

In both types of boiler, the external downcomers consist of tubes about 100 mm in diameter fitted in the double casing of the boiler.

Refractory material is used on the furnace floor and front burner wall in both types of boiler. It is also used behind the water walls, etc. In addition the Babcock and Wilcox boiler also uses studded tube walls.

The combustion air passes through ducting arranged in the double casing of the boiler, and it then passes through air registers into the furnace where combustion takes place.

Internal access to tubes, etc. is obtained by manholes in the steam and water drums, and by handholes in the water wall headers.

In the later versions of these boilers the double casing is all welded.

All the tubes, including the downcomers, are expanded and bell-mouthed into drums and headers.

The drums are of all-welded construction, the water wall headers are solid forged, with welded ends. All the tubes are solid-drawn mild steel.

Q. Sketch and describe a controlled superheat boiler of a two furnace type; give tube sizes and show the position of the superheater. Explain how the superheat temperature is controlled. Indicate gas and water flows for the boiler.

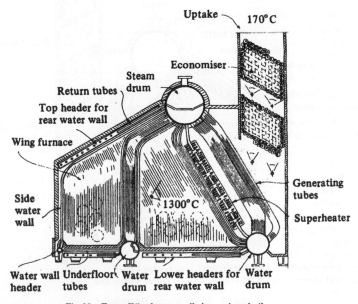

Fig. 32   Foster Wheeler controlled superheat boiler

A. For operational superheat temperature above 430°C some means of temperature control becomes necessary. One method of doing this is to have two furnaces; no dampers or attemperators are fitted, the temperature being controlled by the rate of firing in the individual furnaces. The type of controlled superheat boiler illustrated in Fig. 29 has one steam drum, two water drums, and two furnaces, the latter being separated by an intermediate bank of generating tubes.

Both furnaces have water walls at roof, sides, and rear, with refractory material on floors and front burner walls.

External downcomers supply the water drums, while the lower headers for the water walls are supplied by underfloor tubes in a similar manner to the D-type boiler.

The water wall tubes and the two rows of screen tubes adjacent to the furnaces in each of the tube banks consist of 50 mm diameter tubes. The generating tubes themselves are 32 mm in diameter, and the external downcomers 100 mm in diameter.

By burning a greater proportion of fuel in the wing furnace the superheat temperature is reduced as, although the total mass of exhaust gases flowing over the superheater remains the same, its overall temperature has been reduced due to the greater proportion of these gases which have passed over the intermediate tube bank placed between the furnaces. This type of boiler gives a control range of about 60°C over the final superheat temperature.

The wing furnace is used to raise steam, thus giving added protection to the superheater.

Q. Sketch and describe a selectable superheat boiler. Give tube sizes, and show the position of the superheater. Explain how the superheat temperature is controlled. Indicate the gas and water flows for the boiler.

A. This boiler is similar in basic layout to the normal type of two drum, bent tube boiler, except for the divided gas paths from the furnace. It has two drums. The larger steam drum is placed above a smaller water drum, the two being connected by numerous generating tubes.

The furnace is placed to one side, and the hot gases leaving the furnace pass over three rows of 50 mm diameter screen tubes to enter two passages placed in parallel, and divided by a studded tube wall. The superheater is placed in one of these passages, the gas flow across it being controlled by linked dampers which, while allowing the total mass flow of gas to remain constant, regulates the proportion of gas flowing through the individual passages.

The dampers can be controlled by either manual or automatic means, so giving a wide range of control over the final superheat temperature under all steaming conditions.

The superheater only reaches across one gas passage and, in order to fulfil the required surface area, it is of W-form.

A bank of 32 mm diameter generating tubes is placed behind the superheater, while in the saturated passage a bank of 38 mm diameter tubes replaces the superheater.

After leaving the generating tubes the gases enter the damper section. The gas streams then combine and enter economisers, etc.

External downcomers, 100 mm in diameter, fitted in the double casing supply water from the steam drum to the water drum, and the lower headers for the water walls.

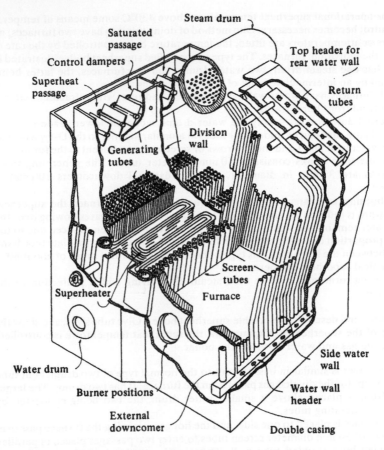

Fig. 33 Babcock and Wilcox selectable superheat boiler

Water walls are fitted to the roof, side, and rear wall of the furnace. These walls consist of 50 mm diameter tubes, partially studded to provide anchorage for refractory material used to seal the space between the tubes. Refractory material is also used on the floor and front burner wall of the furnace.

The steam drum is fitted with cyclone steam separators, and must be arranged in a fore-and-aft position.

The fuel oil burners are grouped in the front wall of the furnace. Combustion air flows through ducting arranged in the double casing to pass through the air registers into the furnace where combustion takes place.

Internal access to the boiler is obtained by manholes in the steam and water drums, and by handhole doors in the headers.

The drums are of all-welded construction; the water wall headers are solid forged, with welded ends. All the tubes are solid drawn.

Q. Sketch and describe an ESD.1 type boiler. Give tube sizes and show the position of the superheater. Indicate the gas and water flows, and give gas temperatures from furnace to funnel.

A. The basic design consists of two drums, the larger steam drum being placed above a smaller water drum, with a water cooled furnace placed to one side in a manner similar to that of the two drum, bent tube boilers. See Fig. 34.

The furnace has water walls consisting of close-pitched 50 mm diameter tubes at roof, side, and rear. The lower headers for these water walls are of rectangular section, and are supplied with water from the water drum by means of underfloor tubes. External downcomers, 130 mm in diameter, supply the water drum.

The gases leaving the furnace pass over eight rows of 50 mm diameter screen tubes before reaching a multi-loop type superheater placed in the lower part of the uptake. This superheater consists of mild steel elements, expanded and bell-mouthed into the headers, up to a temperature of 455°C—above this temperature alloy steel tubes welded to alloy steel headers are used. These superheater elements are suspended from beams cast from heat-resistant steel, and supported at their inboard ends by water cooled support tubes.

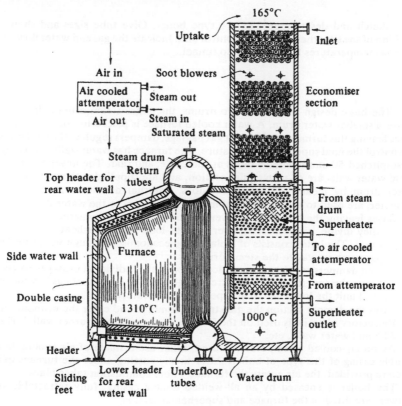

Fig. 34   Foster Wheeler ESD. 1 boiler

39

The gas flow is directed to the bottom of the superheater by a gas baffle consisting of flanged plates, with a layer of air castable refractory on the furnace side.

Control over the final superheat is achieved by means of an air cooled attemperator placed in circuit between the primary and secondary superheaters.

After leaving the superheater the gases enter the economiser section so heating the incoming feed water.

Refractory material is used on the furnace floor and front burner wall. It is also used behind the water walls, etc.

After leaving the attemperator, the heated combustion air passes through the air registers into the furnace where combustion takes place.

Internal access to tubes, etc. is obtained by means of manholes in steam and water drums, and by handholes in the water wall headers.

The drums are of all-welded construction; the headers are solid forged, with welded ends. All the tubes are solid drawn.

The boiler is enclosed by an all-welded double casing.

Q. Sketch and describe an ESD.11 type boiler. Give tube sizes and show the position of the superheater and the control unit. Indicate the gas and water flows, and give gas temperatures from furnace to funnel.

A. The basic design consists of two drums, the larger steam drum being placed above a smaller water drum. A water cooled furnace is placed to one side, and the gases leaving this furnace enter a split gas passage, dampers regulate the gas flow and so control the final superheat temperature. The furnace has water walls consisting of close-pitched 50 mm diameter tubes at roof, side, and rear. The lower headers for these water walls are of rectangular section, and are supplied with water from the water drum by means of underfloor tubes. External downcomers, 130 mm in diameter, are used to supply water from the steam drum to the water drum.

Gases leaving the furnace pass over six rows of 50 mm diameter screen tubes, where they divide, either passing over a multi-loop type superheater or over the control unit. The latter consists of a plain tube economiser through which the feed water passes on its way to the steam drum.

Linked dampers regulate the gas flow through the individual passages so giving control over the superheat temperature; the greater the amount of gas flowing over the control unit the lower the final superheat temperature.

After the damper section the gas streams combine and enter the economiser.

Refractory material is used on the furnace floor, and front burner wall. It is also used behind water walls, etc.

As no air-cooled attemperator is used, fuller use is made of air cooling in the double casing of this type of boiler. After finally passing under the furnace in the ducting provided, the air enters the furnace where combustion takes place.

The boiler is encased by an all-welded double casing; fully retractable soot blowers are fitted in the furnace and superheater section.

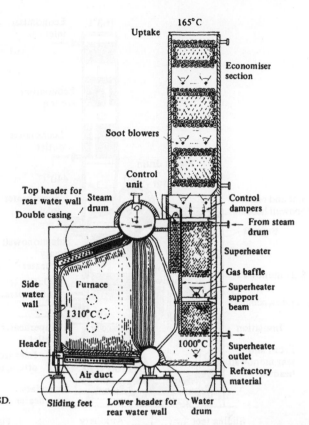

Fig. 35 Foster Wheeler ESD. 11 boiler

Q. Sketch and describe an ESD.111 type boiler. Give tube sizes and show the position of the superheater. Describe the method of superheat control used in this type of boiler. Indicate gas and water flows, and give gas temperatures from furnace to funnel.

A. This boiler has two drums, and a fully water-cooled furnace with the fuel oil burners mounted in the roof. Control over the final superheat temperature is achieved by means of a water-cooled attemperator mounted in the steam drum. Some versions of this boiler have water walls of the close-pitched type while others use monowalls. This latter type of water wall consisting of finned tubes welded together along their length enables a single casing to be used in place of the double casing required with the more orthodox form of water wall.

A dividing water wall completely separates the upper portion of the furnace from the superheater section; only in the lower part of the wall are the tubes splayed out to enable the gases from the furnace to pass through on their way to the multi-loop type superheater.

The superheater consists of primary and secondary sections, a water cooled attemperator mounted in the steam drum and placed in circuit between them giving

41

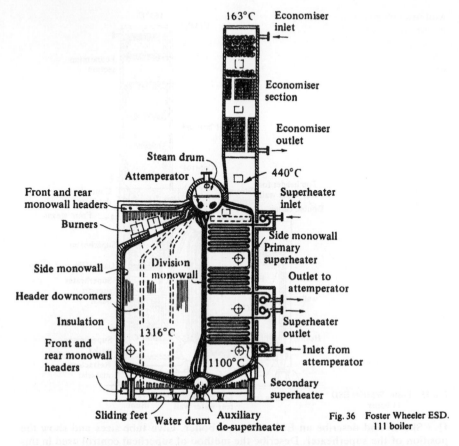

Fig. 36 Foster Wheeler ESD. 111 boiler

control over the the final superheat temperature. A control valve is used to regulate the amount of steam actually passing through the attemperator, the remaining steam going directly from the primary to the secondary superheater.

It should be noted that parallel steam flow is used in the secondary superheater in order to prevent the hottest steam and the hottest gas coming into contact. This is done to reduce the amount of bonded deposits forming on the superheater tubes.

The superheater tubes are welded to the superheater headers, and suspended from support beams made of heat resistant steel. The tube materials are graded to suit the temperature conditions to which they are exposed. Mild steel tubes give way to alloy steel tubes as the steam temperature increases.

The furnace is designed for roof firing; this allows for a long flame path which gives a more even release of heat and allows a longer time for combustion to take place, thus allowing very small amounts of excess air to be used. However, with the burners mounted on top of the boiler remote from normal control positions, the use of automatic combustion control is essential on this type of boiler.

External downcomers are used to supply both the water drum and the lower water wall headers. The diameter of the water wall tubes depends upon the form of wall used; the close pitched use 50 mm diameter tubes, while the monowall type uses 62 mm diameter. Water walls are also fitted to the superheater section.

After leaving the primary superheater the gases enter the economiser, and in some cases if the feed temperature into the economiser is high enough a gas/air heater may then be fitted.

Both steam and water drums are of all-welded construction; all the tubes are solid drawn.

Fully retractable sootblowers are fitted in way of the superheater section.

Q. Sketch and describe a water tube boiler using tangential firing in the furnace.

A. The basic layout shown in Fig. 37 consists of two drums, the larger steam drum being arranged above the smaller water drum, with a bank of 32 mm diameter generating tubes connecting them. The furnace is placed to one side, with the hot combustion gases passing over two rows of 64 mm diameter screen tubes to enter the superheater

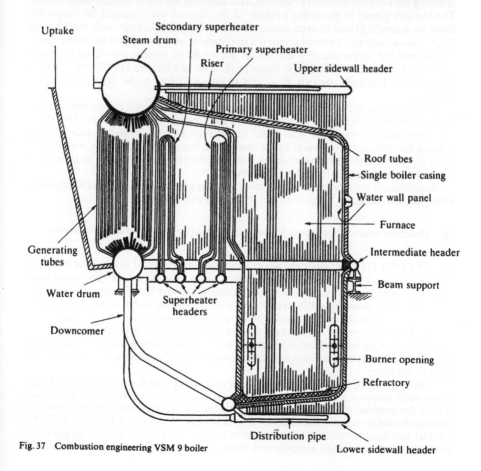

Fig. 37   Combustion engineering VSM 9 boiler

placed in front of the generating tube bank. The superheaters are mounted vertically to give good drainage, and to avoid horizontal surfaces onto which slag deposits can build up.

To avoid the formation of bonded deposits on the gas surfaces of the superheater tubes, the metal temperatures are reduced by preventing the hottest gas and steam flows coming together. The hottest gases first enter the primary superheater and its lower steam temperatures, before flowing on to the secondary superheater with its higher steam temperatures. An attemperator mounted in the steam flow line between the two superheater sections provides control over the final superheat outlet temperature.

The boiler is supported at mid-height by a saddle beneath the water drum, and by supports under the intermediate waterwall headers placed level with the water drum. The furnace lies partly above and partly below these intermediate headers, so being able to expand both upwards and downwards from the point of support.

Gastight diaphragm walls of welded construction completely surround the furnace, reducing the need for refractory material, and allowing the use of a single boiler casing. The burners placed in the lower portion of the furnace are arranged for tangential firing by mounting them in each corner of the rectangular furnace, with each burner aligned tangential to a circle of about 0·6 m diameter in the centre of the firing space. This allows the formation of a horizontal vortex flame pattern providing increased turbulence and an effectively longer flame path, both of which lead to efficient combustion with low levels of excess air, so greatly reducing corrosion and deposits in the gas passages.

Positive water circulation in the boiler is provided by external downcomers which first feed the water drum and then from there the lower water wall headers. This water then rises through the water wall tubes to the intermediate headers which serve to balance the distribution of water to the upper sections of the water walls, where risers from the top headers to the steam drum complete the circulation path. The generating bank is supplied directly from the water drum.

Economisers are mounted in the uptake after the generating tubes, and if required a rotary air heater can be fitted as a final stage heat exchanger.

Q. Describe some of the design features of radiant heat type water tube boilers.

A. The ESD.111 type boiler shown in Fig. 36 forms a typical example of a radiant heat boiler. These have no banks of generating tubes, relying instead upon water walls which completely surround the furnace, receiving radiant heat for the generation of steam. In a few cases tangential type water walls are fitted, but in general diaphragm walls are preferred; these form a water box type furnace and require only a single boiler casing. Many versions of radiant heat boilers use welded tube attachment, and in this case only a single boiler drum is required, with a header being fitted in place of the water drum. This is because internal access is not required in order to expand the lower ends of the tubes.

The majority of radiant heat boilers use roof firing which, by providing a longer flame path, allows a greater time period for the combustion process to be completed within the confines of the furnace. This provides for more efficient combustion with reduced excess air requirements, below 5% as compared to the 5% to 15% needed for normal side-fired furnaces. As a result there is a less intense but more even release of heat, avoiding local high temperature regions which could be detrimental to the diaphragm type water walls, and also allowing a more even distribution of the gas flow across the superheater.

This superheater is contained in a water cooled space and comprises primary and secondary multi-loop type superheaters, with a water cooled surface or spray type attemperator placed in circuit between them to give control over the superheat temperature. Parallel flow conditions are often used in the superheater sections in an attempt to reduce the formation of bonded deposits by preventing the hottest steam and gas flows coming together. In most cases the arrangement consists of a counter flow primary, with a parallel flow secondary unit placed beneath it. The superheat materials are graded to suit the steam temperatures involved.

After the superheater, an economiser is fitted to provide the vital heat exchange surface needed to raise the incoming feed close to the evaporation temperature.

In some cases, to avoid circulation problems, a steaming economiser is fitted; these have parallel flow so that any steam bubbles forming can pass upwards through the tube elements to enter the steam drum, and not interfere with the feed flow.

Finally, to achieve maximum boiler efficiency a rotary air heater may be fitted after the economiser section.

Q. Describe the methods of tube attachment used in water tube boilers.

A. Tubes can be attached to drums and headers by expanding or by welding. In most cases the generating, screen, and water wall tubes are expanded into plain seats, and then bell-mouthed. The tube ends must be cleaned, and then carefully drifted, or roller expanded into the holes in the tube plate. They must project through the tube plate by at least 6 mm. To prevent tubes pulling out of the tube plate, they must be bell-mouthed. This bell-mouthing is to be 1 mm for every 25 mm of outside diameter plus 1·5 mm. See Fig. 38.

When the tube enters perpendicular to the surface of the tube plate, the length of parallel seating must be at least 10 mm. Where the tube does not enter perpendicular to the surface, the length of parallel seating must be at least 13 mm.

In the case of tubes with larger diameters, such as downcomers, it is usual to use grooved seats. The tube material flows into the grooves during the expansion process so helping to form a tight seal.

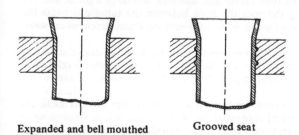

Expanded and bell mouthed   Grooved seat   Fig. 38 Expanded tube attachment

Superheater tubes are also usually expanded and bell-mouthed up to steam temperatures of about 450°C; above this value the tubes are often attached by welding. As can be seen in Fig. 39 a stub tube is interposed between the actual tube and the header. Two methods of welding are shown. In the gas-welded tube a

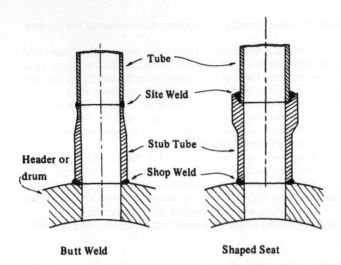

Butt Weld   Shaped Seat

Fig. 39   Welded tube attachment

backing ring may be required to prevent weld metal breaking through into the tube bore. In the other type, backing rings are not necessary due to the specially shaped tube end. Electric welding may be used. However, the tube must make a ninety degree bend close to the joint in order to prevent direct tensile stresses coming onto the weld.

Q.   Describe with the aid of sketches the basic designs of water walls used in the construction of water tube boilers.

A.   Water walls are used in water tube boilers to contain the heat of the furnace and reduce the amount of refractory material required. They also contribute to the evaporation rate by receiving radiant heat; indeed, in radiant heat boilers the water walls provide the bulk of steam generation.

As a design factor, the tube pitch is critical: too small, and the corresponding attachment holes unduly weaken tube plates and headers; too large a pitch, and the amount of radiant heat passing through the gaps between the tubes overheats the boiler casing. Four basic designs of water walls to overcome these problems are illustrated in Fig. 40.

The partially studded water wall has a relatively wide pitch and so does not unduly weaken the tube plates or headers, but to seal the gaps between the tubes a refractory material must be used, this being keyed in place by steel studs resistance-welded to the water wall tubes.

In the case of the tangential type water wall a much smaller pitch is used, the off-set tube ends allowing the use of staggered holes to prevent undue weakening of tube plates and headers. However, some radiant heat can still penetrate the wall and a layer of suitable refactory material must be placed behind the water to protect the boiler casing.

Diaphragm type water walls are constructed by welding longitudinal fins between adjacent plain tubes, as in the membrane wall, or alternatively welding finned tubes together along the line of contact, as in the monowall. In both cases similar gastight

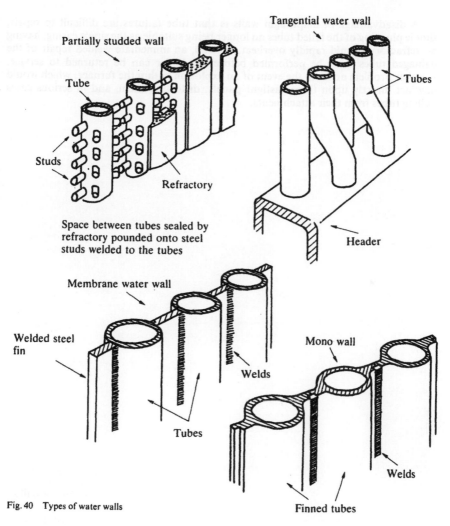

Fig. 40 Types of water walls

panels are formed which can be used to surround the furnace, the welded fins being omitted in way of any required gas passages. As no radiant heat can penetrate these walls, no refractory material need be fitted behind them and a single boiler casing can be used.

Again tube pitch remains a critical factor; too narrow a pitch not only unduly weakens the tube plates, but can also overcool the furnace flame leading to unstable combustion conditions. Too wide a pitch can result in overheating of the longitudinal fins.

The intense short flames suitable for use with tangential or partially studded water walls are not suitable for diaphragm water walls, which need a low furnace heat flux such as that produced by a long flame in conjunction with low values of excess air

47

A disadvantage of diaphragm walls is that tube failures are difficult to repair, simple plugging of the failed tubes no longer being suitable as the single casing, having no refractory, would rapidly overheat. Instead, an immediate welded repair of the damaged tubes must be performed before the boiler can be returned to service. Another problem arises in the event of an explosion within the furnace, which would now act directly upon these gastight panels, distorting them, and in serious cases pulling tubes from their attachments.

CHAPTER 4

Superheaters and Uptake Heat Exchangers

Q. Discuss the reasons for the use of higher pressures and temperatures in modern steam plant.

A. In general for any heat engine cycle the following statement can be made regarding its thermal efficiency:

Thermal efficiency is proportional to the ratio

$$\frac{T_1 - T_2}{T_1}$$

Where

T_1 = The highest absolute temperature in the cycle.
T_2 = The lowest absolute temperature in the cycle.

Thus to increase the thermal efficiency the temperature difference $(T_1 - T_2)$ must be increased.

In a steam plant the two temperatures to be considered are the steam supply temperature (T_1), and the exhaust temperature (T_2).

To understand the factors governing these two temperatures, consider a unit mass of water to be heated at constant pressure. The process is shown in diagrammatic form in Fig. 41.

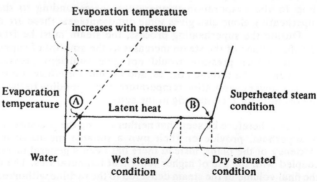

Fig. 41

At first the temperature of the water increases and this continues until it reaches its evaporation temperature, indicated by point A on the diagram. It then absorbs latent heat, the temperature remaining constant as indicated by the horizontal line on the diagram. This process continues until all the water present has changed to steam;

49

in this condition, indicated by point B on the diagram, it is referred to as dry saturated steam. At this point if more heat is added, still at constant pressure, the temperature of the steam increases, and it is said to be in a superheated condition. If, on the other hand, heat energy is removed from steam at the dry saturated condition, it will begin to condense, the resulting mixture of steam and water being referred to as wet steam.

From the diagram it can also be seen that as pressure increases, the corresponding evaporation temperature increases, while the latent heat of evaporation decreases.

Thus the exhaust temperature is governed by the exhaust pressure, and in a steam turbine using a regenerative condenser, the exhaust pressure is close to absolute zero pressure, so there is little possibility of further improvement here. Therefore the only practicable way to increase the ratio is to increase the highest absolute temperature (T_1) in the cycle. There are two basic ways in which this may be accomplished:

RAISING THE BOILER PRESSURE

This increases the evaporation temperature of the steam, but gives the following disadvantages:

To keep the stresses in the boiler material within the desired values, the boiler scantlings must increase as boiler pressure increases.

As seen on the diagram, when dry saturated steam gives up energy at constant pressure, it begins to condense. The water droplets thus formed lead to increased friction, and also to erosion, as the steam passes over the turbine blades.

From the steam tables, it can be ascertained that little increase of the energy stored in the steam occurs over a wide range of pressure increase. This is due to the fact that, although the sensible heat energy increases as the evaporation temperature increases, it is off-set by the corresponding decrease in the latent heat value.

SUPERHEATING THE STEAM

As shown on the diagram, if dry saturated steam is further heated at constant pressure, its temperature increases, and thus it stores more energy without increasing the boiler scantlings. The onset of condensation is also delayed, the steam having to drop to the evaporation temperature corresponding to the pressure. However superheating alone also gives some disadvantages; these are as follows:

During the superheating process, the steam must be free to expand, thus the specific volume of the steam increases as the amount of superheat increases.

Low boiler pressure would not give sufficient pressure drop through the superheater, the turbine nozzles, and turbine for efficient operation.

The lower evaporation temperature corresponding to the low pressure limits the opportunity for feed heating in the system.

It can therefore be seen that neither raising the pressure, nor increased amounts of superheat, provide on their own a suitable means of increasing the thermal efficiency of the plant; instead there has been a general increase in boiler pressure coupled with the use of higher superheat temperatures. The higher pressure keeps the final volume of the steam delivered to the turbine within reasonable limits, giving the higher evaporation temperatures required for the fuller use of feed heating. Finally the higher pressure gives the necessary pressure drop required for the proper steam flow through the superheater, etc. It also enables the pressure drop through the turbine to be arranged so that the steam temperature at any stage is always above the corresponding evaporation temperature; condensation is therefore prevented from taking place until the final rows of the low pressure turbine.

Q. Discuss the need for control over superheat temperatures in water tube boilers. State methods by which this control can be achieved.

A. In a water tube boiler, the water flowing through the tubes conducts the heat away so rapidly that the tube metal temperatures lie much closer to the water temperature than to the gas temperature. Steam, however, due to its lower specific heat cannot conduct heat away from the tube metal so quickly, therefore superheater tubes tend to work at higher metal temperatures than those containing mainly water, e.g. screen, water wall, and generating tubes. However, for superheat temperatures up to about 455°C, no serious metallurgical problems arise and ordinary mild steel tubes can be used. For superheat temperatures above this value, alloy steels containing small amounts of chrome and molybdenum are necessary in quantities graded to suit the operating conditions. This greatly increases the cost of the superheater, and also presents difficulties in welding the tubes; this is necessary as, at these elevated temperatures, expanding the tubes into the headers proves unsatisfactory. However, even with the use of alloy steels the temperatures must be kept within certain limits, and at the present time superheat temperatures in the order of 560°C form the upper limits for marine use.

Thus it is necessary to provide some form of superheat control in order to keep the temperature below 455°C so that mild steel tubes can be used, or within the limits required for the alloy steel being used.

When the operating superheat temperature is well below the limit for mild steel, sufficient control can be achieved by the position of the superheater within the boiler where, other factors being equal, the more rows of tubes placed between the furnace and the superheater, the lower will be the steam temperature.

With this form of control fluctuations of temperature, caused by changes in load, can only be kept in check by the rate of manoeuvring. This means that as operating temperatures are increased undue limitations are imposed upon the manoeuvring rate and so some further means of control become necessary. This can be achieved by various methods:

By using dampers to control the flow of gases over the superheater.

Designing the boiler with two separate furnaces so that the superheat temperature can be controlled by the amount of fuel burnt in the individual furnaces.

By the use of an attemperator. These can be air or water cooled, and are fitted in circuit between primary and secondary superheaters.

Another method, occasionally used where the superheater is well protected from direct radiant heat, is to allow some of the steam to make a reduced number of passes through the superheater by means of a by-pass valve fitted to the superheater header.

All of these methods are used basically to protect the superheater; i.e. by preventing the metal temperatures from reaching too high a level.

Another form of control, known as de-superheating, is necessary to provide lower temperature steam for use in auxiliary machinery, or where limitations are imposed on the steam temperature to be used in an astern turbine. Two main forms of de-superheater are in general use; one is a surface type, similar in construction to an attemperator. The other consists of a water spray, either mounted in a separate

51

chamber, or directly in the steam line. The latter is used in a few cases to control the steam temperature entering a turbine casing.

The reason for using a de-superheater, instead of taking saturated steam directly from the boiler, is to give additional protection to the superheater by providing some circulation of superheated steam—even when the turbine throttle is closed—so preventing the superheater tubes from overheating.

A final reason for the control of superheat temperatures is that in the combustion of residual oil, certain impurities (mainly sodium and vanadium salts) cause bonded deposits to form on the superheater tubes. One way of reducing these deposits is to prevent the hot gases which leave the furnace from coming into direct contact with the superheater. This is done by placing rows of screen tubes to reduce the gas temperature, and in many modern boilers by placing the superheater remote from the furnace. In addition with multi-loop superheaters using high superheat temperatures, the final superheat section uses parallel flow, so preventing the hottest gas and steam temperatures coming into contact with each other.

Q. Sketch and describe a superheater as fitted to a header type water tube boiler. State the materials used, and tube sizes.

A. The superheaters fitted to header type water tube boilers consist of a series of U-tubes, expanded and bell-mouthed into two separate cylindrical, or rectangular section, forged steel headers. These are placed out of the gas stream, to one side of

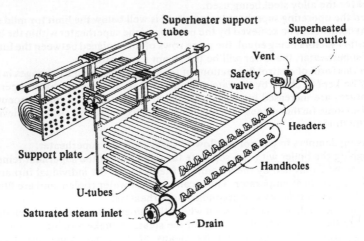

Fig. 42 Superheater in a header boiler

the boiler, in a plane parallel to the inclined generating tube bank. The superheater tubes project horizontally across the gas stream, being placed at right angles to the generating tubes.

Internal baffles are welded in place within the headers so that the steam makes the required number of passes through the gas stream.

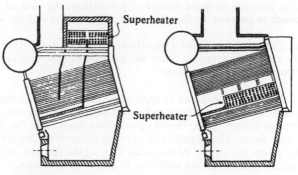

Fig. 43 Alternative superheater positions in header
boilers

Three pass boiler
low temperature superheater

Single pass boiler
high temperature superheater

For low superheat conditions the superheater is placed above the return tubes, at the top of the first gas pass. For higher steam temperatures it is placed in an interdeck position, above two rows of 50 mm diameter screen tubes, and then as many rows of 32 mm diameter generating tubes as necessary to give the required superheat temperature. The width of the superheater depends upon the quantity of steam produced; in the larger boilers it will reach right across the available space.

In both cases the superheat temperatures will be below 455°C, so that ordinary solid-drawn mild steel tubes can be used.

The superheater tubes, 32 mm to 38 mm in diameter, are supported by means of transverse support plates which, in the case of the interdeck type, are carried by superheat support tubes.

Apart from the basic control over steam temperature imposed by the position of the superheater within the boiler, some control can be exercised over the final superheat temperature by fitting control dampers in the gas baffles to regulate the gas flow over the superheater. Alternatively by-pass valves, fitted to the superheater headers, allow some of the steam to by-pass the internal baffles in the headers, and so make fewer passes through the superheater.

The superheaters must be protected from overheating whenever hot gases are flowing across them, and so provision must be made for circulation of steam. Especial care is required at low load conditions, and de-superheaters are often fitted in order to provide sufficient steam flow through the superheater under these conditions. Another precaution is that the safety valve fitted on the superheater outlet header is set to lift at a lower pressure than the drum safety valves; this ensures steam flow through the superheater in the event of the safety valves blowing off.

When raising steam, protection is given in some cases by means of control dampers which reduce the gas flow over the superheater; while in the case of interdeck superheaters, some can be partially filled with water through special flooding valves before flashing up. When sufficient steam has been produced to circulate the superheater to atmosphere, this water can be drained out. The disadvantage with this method of protecting the superheater is the possibility that any solids in the water can deposit out in the superheater tubes before the draining takes place.

In addition to these means of protection, the rate of firing must be kept low enough to prevent overheating taking place in way of the superheater.

Q. Sketch and describe a superheater as fitted to a D-type water tube boiler. State tube sizes and materials. Show the method of tube support. How is the superheater protected from overheating?

A. Integral furnace and D-type water tube boilers usually have their superheaters placed between the steam and water drums, behind two or three rows of 50 mm diameter screen tubes, thereby protecting them from the direct radiant heat of the furnace flame. The superheater consists of series of U-tubes which project horizontally across the stream of hot gases leaving the furnace. In many cases the tubes are given a slight slope, just sufficient to allow for drainage.

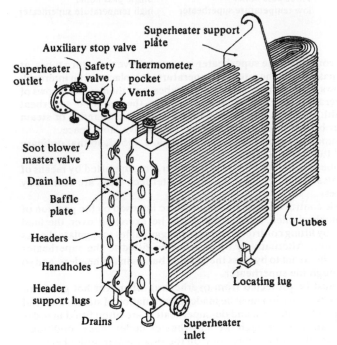

Fig. 44 Superheater in a D-type boiler

As these boilers operate with superheat temperatures below 455°C, solid-drawn mild steel tubes can be used. These tubes, 32 mm in diameter, are expanded and bell-mouthed into two separate, cylindrical or rectangular section, forged steel headers which are arranged in a vertical position, outside the gas stream, at the rear of the boiler.

Internal baffles, welded inside these headers, cause the steam on its way through the superheater to make several passes across the gas stream. Small holes in these baffle plates allow for drainage. Internal access to the headers is provided by means of hand holes; these are closed by plugs kept in position by external dogs.

54

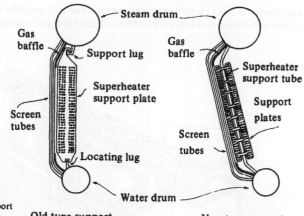

Fig. 45 Superheater tube support plates

Old type support   New type support

The superheater tubes are supported by plates made of heat-resistant alloy steel. In some versions these plates are suspended from lugs welded to the underside of the steam drum, with a locating slot, and plate at the bottom as shown in Fig. 43. Other designs use a water-cooled support tube, fitted between the steam and water drums, to carry a number of support plates which are clamped onto it.

Gas baffles are fitted to direct the gas flow over the superheater as required.

No special means of control over the superheat temperature is provided as the normal operating temperatures are sufficiently below the limit for mild steel to allow some temperature fluctuations to occur while manoeuvring, etc. However, care must be taken with rate of firing at these times to prevent excessive variation in superheat temperatures.

As with all superheaters, arrangements must be made to prevent overheating by allowing for the circulation of steam at all times when hot gases are flowing over the superheater. Thus some steam-driven auxiliary machinery, such as turbo-feed pumps, are advisable to ensure a sufficient minimum circulation when the main engine throttle is closed. De-superheaters are also usually fitted.

The safety valve fitted to the superheater outlet header must be set at a pressure sufficiently below that of the drum safety valves to ensure that it opens first and so provides steam flow through the superheater in the event of the boiler blowing off.

Due to the risk of internal deposits, these superheaters are not flooded when raising steam. Drains and air vents are left open, and the rate of firing kept low, until enough steam is being generated to circulate the superheater; the steam is allowed to blow off to the atmosphere until some steam demand exists to provide circulation through the superheater.

Q. Sketch and describe the superheater fitted to a selectable superheat water tube boiler. State the tube sizes, and show the method used to support them. How is the superheater protected from overheating?

A. As operating steam temperatures come near the limit of 455°C for mild steel, little variation in superheat temperatures can be allowed while manoeuvring, etc.

and some form of superheat temperature control must be provided. The selectable superheat boiler does this by means of divided gas passages and control dampers.

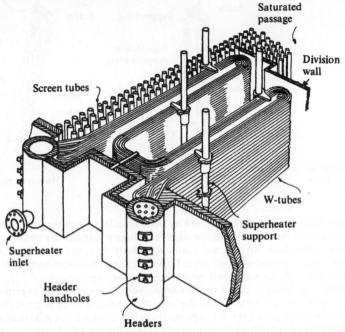

Fig. 46   Superheater in a selectable superheat boiler

The superheater in this type of boiler is placed between the steam and water drums, and consists of a series of W-shaped tubes projecting horizontally across the superheater gas passage. This form of tube is used for two reasons; one is that, due to the saturated gas passage, the superheater cannot reach right across the full width of the gas stream leaving the furnace. The other is that due to the problem of bonded deposits forming on the gas side of the superheater the tubes are given a wider pitch, so reducing the possibility of these deposits building up, bridging the gap between the tubes, and so choking the gas passages through the superheater. Thus the restriction in length, and the wider pitch, lead to the need for the W-form of tube to provide the required heat exchange surface.

The solid-drawn mild steel tubes, 32 mm in diameter, are usually attached to the headers by expanding and bell-mouthing, although in a few cases welding may be used.

The forged-steel cylindrical headers are placed outside the gas stream in a vertical position at the front of the boiler. Internal baffles welded inside these headers cause the steam to make several passes across the gas stream. Handholes provide internal access to the headers.

Because of the wider pitch of the tubes, and the higher operating temperatures, support plates are no longer suitable and finger plates, clamped onto water cooled support tubes, as shown in Fig. 47, are used instead.

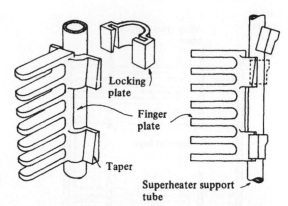

Fig. 47 Finger plate type tube supports

Locking plate

Finger plate

Taper

Superheater support tube

Three rows of 50 mm diameter screen tubes are used to protect the superheater from direct radiant heat. Gas baffles are fitted to direct the flow of gases across the superheater as required.

The two separate gas passages and control dampers, as well as providing control during load changes, also give protection during flashing up since the gases are allowed to flow through the saturated passage until sufficient steam is available to circulate the superheater.

The superheater safety valve is set at a pressure sufficiently below the drum safety valves to ensure it lifts first and so provides a flow of steam through the superheater if the boiler blows off.

Access space is provided on the gas side of the superheater to give easy entry for cleaning and maintenance.

Q. Sketch and describe a multi-loop type superheater.

A. With increased evaporation rates and higher superheat temperatures, many modern boilers use multi-loop type superheaters. These superheaters are placed remote from the furnace in the lower part of the uptake and consist of a number of multi-loop tube elements, expanded and bell-mouthed, or more usually welded to the headers, which in this case contain no internal baffles; the steam enters all the tubes together, but makes several passes across the gas stream via each element. See Fig. 48.

This form of superheater provides more flexibility in layout, and enables sufficient heat exchange surface to be provided without unduly cramping the space available for the superheater. Better access for cleaning and maintenance is therefore provided, and the possibility of a build-up of deposits on the gas side, which would choke the gas passages through the superheater is reduced.

Multi-loop superheaters usually operate at temperatures above 455°C and so require the use of alloy steels containing small amounts of chrome and molybdenum. The tube materials are graded, starting with solid-drawn mild steel and then, as the steam temperatures increase, changing to an alloy steel via a welded transition piece.

The headers are placed outside the gas stream in a horizontal position. The inlet header is made of mild steel while, if the final superheat temperature is above 455°C, the outlet header is of alloy steel.

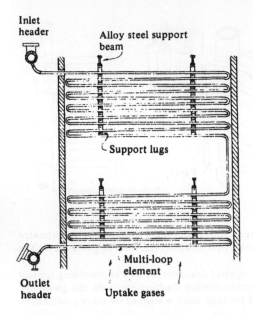

Inlet header

Alloy steel support beam

Support lugs

Multi-loop element

Outlet header

Uptake gases

Fig. 48   Multi-loop type superheater

The superheater tubes are supported by a series of lugs made from heat resistant alloy steel, fitted between each U-section, as shown in Fig. 48. These lugs in turn are supported by means of beams of heat resisting alloy, placed across the gas passage, above the element being supported. As an alternative, the top tube section in each element has a steel rib welded along its length so that it forms a water cooled support beam.

In both cases the inboard ends of these beams are carried by vertical water cooled support tubes fitted between the steam and water drums.

Provision must be made for the circulation of steam under all conditions of load, and invariably some form of superheat temperature control is fitted.

The safety valve placed on the superheater outlet header is set at a pressure sufficiently below that of the drum safety valves to ensure a flow of steam through the superheater in the event of the boiler blowing off.

Precautions while raising steam vary with the type of superheat control fitted, but vents and drains should always be left open until enough steam can be circulated to protect the superheater.

Q.   Sketch and describe both a water, and an air cooled attemperator. Explain the purpose for fitting these to a water tube boiler.

A.   Attemperators are fitted in order to keep the steam temperature in the final stages of the superheater within certain desired limits.

This is done for two main reasons. The most important is the protection of the superheater tube material from the effects of overheating, as it is the steam temperature which governs the tube metal surface temperature rather than the temperature of the gas flowing over them. Thus, for example, mild steel superheater tubes should not be subjected to steam temperatures in excess of 455°C.

58

It has also been found that the formation of bonded deposits on superheater tubes can be reduced if steam and gas streams within certain temperature ranges are prevented from coming into contact with each other.

Two basic types of attemperators are available; these are described as follows:

WATER COOLED ATTEMPERATOR

These are constructed in a manner similar to that of a surface type de-superheater. They are placed below the water level in the steam drum as shown in Fig. 49 or, as an alternative, placed in an external chamber connected by balance connections to the circulation circuit of the boiler.

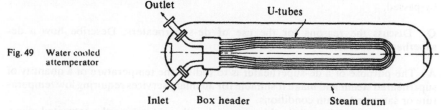

Outlet   U-tubes

Fig. 49   Water cooled
attemperator

Inlet   Box header   Steam drum

With this type of attemperator some of the steam leaving the primary superheater passes directly to the secondary superheater, while the remainder passes through the attemperator before rejoining the rest of the steam to enter the secondary superheater. Control valves, fitted to the steam line leaving the primary superheater, regulate the final superheat temperature by varying the proportion of steam flowing direct to the secondary superheater to that passing to it via the attemperator.

AIR COOLED ATTEMPERATOR

This is usually fitted in the double casing of the boiler, and consists of inlet and outlet headers connected together by a series of multi-loop tube elements, which enable the superheated steam to make several passes across the path of the combustion air flowing to the furnace.

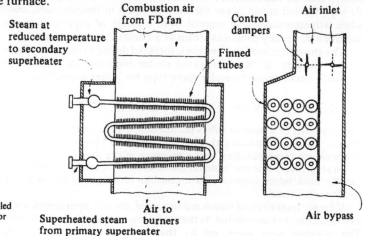

Combustion air
from FD fan

Air inlet

Steam at
reduced temperature
to secondary
superheater

Control
dampers

Finned
tubes

Fig. 50   Air cooled
attemperator

Air to
burners

Air bypass

Superheated steam
from primary superheater

In this type of attemperator, all the steam leaving the primary superheater passes through the attemperator before entering the secondary superheater. The control over the final steam temperature is obtained by regulating the amount of combustion air flowing over the attemperator by means of control dampers and an air by-pass. The general arrangement of this is shown in Fig. 50.

The poor heat exchange efficiency between the air and steam compared to that of water and steam means that air cooled attemperators are much larger than a corresponding water cooled type.

The size of the air cooled version can be reduced to some extent by the use of finned tubes which increase the heat exchange surface.

A steam air heater is usually fitted to heat the combustion air which is being by-passed.

Q. Discuss the reasons for the use of de-superheaters. Describe how a de-superheater operates.

A The purpose of a de-superheater is to reduce the temperature of a quantity of superheated steam and make it suitable for auxiliary services requiring low temperature or saturated steam conditions.

The reason for doing it in this way, rather than by taking the saturated steam direct from the steam drum, is to protect the superheaters; these must be protected against overheating under all conditions of load by circulating steam through them. In this way the steam required for auxiliary purposes is allowed to pass first through the superheater, and is then cooled in a de-superheater. This helps to ensure the necessary circulation for the prevention of overheating, even when the main engine throttle is closed.

Although attemperators and de-superheaters are both used to protect the superheater from overheating, they do this in different ways. Attemperators are fitted to control the temperature of the steam passing through, while de-superheaters help to provide the necessary circulation of steam.

Two other important differences between these two items of equipment are firstly, that de-superheaters are fitted after the final superheater stage and the de-superheated steam is not then raised again in temperature, and secondly, that while attemperators give control over a range of superheat temperature, a de-superheater usually produces steam at a constant temperature.

The choice between the two basic forms of de-superheater available generally lies in the quantity of de-superheated steam required. A surface type is used for relatively small amounts, and a spray type for larger quantities.

SURFACE TYPE DE-SUPERHEATER

These consist of a tube or tubes placed below the water level in the steam or water drums. The steam passing through is cooled to about 30°C above the saturation temperature corresponding to the pressure in the drum.

In some cases fins are fitted on the tubes to extend the heat exchange surface, but usually bare tubes formed into a number of loops, or in the form of a series of U-tubes, are used.

If larger quantities of steam are required, the de-superheater may be placed in an external chamber connected to the steam and water drums by balance connections. The sensible heat given up by the superheated steam passing through the de-superheater is used to generate steam which is then returned via the top balance connection to the steam drum.

SPRAY TYPE DE-SUPERHEATER

These are fitted independent of the boiler, and are suitable for producing large amounts of de-superheated steam. An example of a spray type de-superheater is shown in Fig. 51. It consists basically of a mixing tube, water spray nozzle, and a thermostat to provide control.

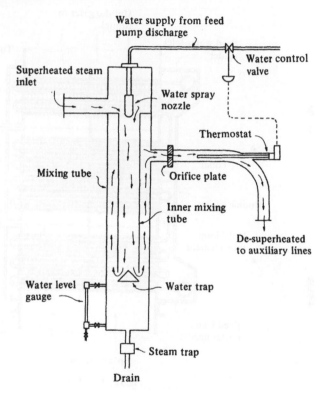

Fig. 51 Spray type de-superheater

Superheated steam enters and passes down through the inner mixing tube. Water is sprayed into this steam through an atomizing nozzle, and an exchange of heat by direct contact takes place. The superheat temperature of the steam is reduced, while the water receiving this heat as latent heat changes to steam. Any water droplets remaining are separated out by the baffle, and are drained off through the steam trap, while the de-superheated steam passes up the outer mixing tube to an orifice plate fitted in the outlet pipe to ensure a thorough mixing and final evaporation of any remaining particles of water before the steam leaves.

The amount of water sprayed in is controlled by means of the thermostat placed in the steam outlet pipe. Thus the steam leaving the de-superheater must be kept a

61

few degrees above the corresponding evaporation temperature to enable efficient control to be maintained.

Q. Sketch and describe an economiser suitable for use with a water tube boiler.

A. Economisers are used to increase boiler efficiency, being placed in the uptakes to lower the temperature of the exhaust gases, and in so doing heat up the incoming feed water as it passes through the economiser on its way to the steam drum.

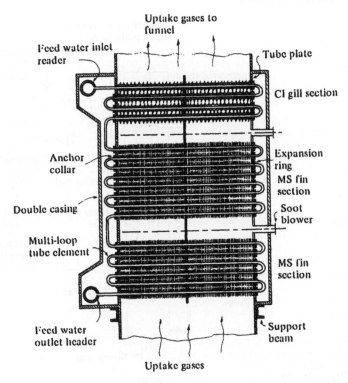

Fig. 52 Economiser

A typical economiser consists of two or more solid-forged mild steel headers, to which are connected series of tubes. These may be attached to the headers either by expanding and bell-mouthing, or by welding, usually by means of stub tubes.

The economiser shown in Fig. 52 consists of a number of multi-loop tube elements enabling the feed water to make several passes across the gas stream as it passes through the economiser.

The materials used for the various economiser sections are governed by the minimum metal surface temperatures required to reduce corrosive attack. These metal temperatures in turn are a function of the feed temperature, being in general some 5°C higher than that of the water passing through the tube.

For feed temperatures above 138°C plain, solid-drawn mild steel tubes may be used, but it is usual to weld mild steel fins or studs onto the tubes so as to extend their heat exchange surface. If the feed water temperature lies between 115°C to 138°C cast-iron gills are used. These completely cover the mild steel tubes, thus protecting them from corrosive attack.

The tubes are supported by vertical tube plates forming the sides of the uptake. In some cases additional intermediate tube support plates are fitted within the uptake.

A relief valve must be fitted to the economiser if it can be isolated from the boiler. In many cases this is not possible since the feed check, and feed regulator are placed on the inlet to the economiser, which then forms an integral part of the boiler.

Air vents must be provided for the release of air when filling the economiser, or when raising steam. Drain valves fitted to the lower headers enable the economiser to be emptied of water when required.

Efficient soot blowing arrangements must be provided so as to keep the heat exchange surfaces free from soot. This is important not only in the interests of efficiency, but also to reduce the possibility of an uptake fire starting in sooty deposits.

Suitable access and the necessary fittings and drains should be provided for the water washing of the uptake heat exchangers when the boiler is shut down.

Q. Discuss with respect to the efficient operation of economisers the importance of maintaining both the gas and water temperatures at their correct values.

A. After leaving the last rows of generating tubes the exhaust gases are still at relatively high temperatures, and heat exchangers placed in the uptakes to reduce this temperature lead to a gain in the overall thermal efficiency of the boiler. However, a practical limitation is imposed upon their use by the fact that if any metal surfaces in the uptakes fall below the 'dew point' temperature at which the water vapour in the exhaust gases condense, then any sulphur dioxide or trioxide present in the exhaust gases, formed by the combustion of any sulphur in the fuel, will be absorbed by the water. This results in the formation of acidic deposits on any metal surfaces below 170°C. This acid can cause corrosive attack on mild steel surfaces at temperatures below 138°C, and on cast iron at temperatures below 115°C.

The feed temperatures are also important as the metal temperatures involved are a function of these feed temperatures, being in general some 5°C higher.

It is thus important to maintain these minimum temperature values under all steaming conditions. If the normal feed temperature entering the economiser is about 140°C there is an allowable margin of some 15–20°C, but the feed temperature should never be allowed to fall below the 115°C limit under steaming conditions; if it does, heavy fouling can occur in addition to the corrosive attack, due to soot deposits building up on the acid dew forming on the cool metal surfaces. Thus some means should be provided to maintain the feed temperatures above these minimum values; for example, by supplying extra steam to a deaerator in a system where one of these is fitted.

These points are vital as the gas temperature leaving the economiser, provided it is above the dew point temperature, has little to do with fouling or corrosion as long as the metal temperatures are kept above the minimum value quoted.

The water temperature within the economiser should not be allowed to exceed a temperature 30°C below the corresponding evaporation temperature; this prevents the formation of steam within the economiser.

63

In many water tube boilers the gas temperatures leaving the last rows of generating tubes are so intense that, due to the high operating temperatures involved, the economiser forms an important part of the heat exchange surface of the boiler. Indeed in some radiant heat type boilers, the gases leaving the superheater immediately enter the economiser and the gas temperature is so high that, in order to prevent the formation of steam in the economiser, the direction of the water flow is reversed. These are referred to as steaming economisers. The water enters at the bottom and then rises parallel to the gas flow. This has the effect of levelling out the temperature gradient and helps to prevent steam forming in the economiser under normal steaming conditions.

Q. Sketch and describe tubular type air heaters as fitted in the uptake of a water tube boiler. Discuss the limitations to the use of gas air heaters.

A. If the uptake temperatures warrant it a gas air heater may be placed in the boiler uptake. However, if this is to be done a critical factor to be considered is that the temperatures of the metal surfaces involved should remain above the dew point temperature of the exhaust gases under all conditions of load, in order to avoid undue deposits and/or corrosion taking place in the uptake. This often prohibits the use of both gas air heaters and economisers at the same time.

In Scotch boilers, and some header type water tube boilers, the gas temperature leaving the last rows of generating tubes is usually so low that only a gas air heater can be fitted. In other types of water tube boilers the gas temperature leaving the generating tubes is higher, and provided sufficient feed heating is available to raise the feed temperature above 140°C economisers are normally fitted. Gas air heaters, placed in the uptake as a final stage of heat extraction after the economiser, can only be considered in a few cases as a viable proposition, i.e. where the feed temperature entering the economiser is above 200°C.

If a choice exists, an economiser will normally be fitted due to the greater efficiency of heat exchange between gas and water, than between gas and air.

When gas air heaters are fitted, those of tubular form are most widely used. These are of two basic types:

VERTICAL TUBE TYPE

These consist of series of vertical tubes, expanded into two horizontal tube plates. In this type the gas passes through the tubes while the air flows over the outside of them. Up to a certain point these are easy to keep clean. However, this cleaning process needs to be carried out manually by means of brushes which are pushed down the tubes when the boiler is shut down. Under steaming conditions soot blowers are used, but are not completely effective because of the tube arrangement. In addition to this, unless extreme care is taken in maintaining the uptake temperatures above their minimum values, heavy deposits can form under low load conditions, often completely choking the vertical tubes; this is especially likely to occur when burning fuels with a relatively high sulphur content. The gas air heater should be by-passed to keep the uptake temperatures as high as possible above their minimum values in order to reduce these deposits.

Under certain conditions these deposits can catch fire, but in the case of vertical tube type heaters these fires tend to be confined to individual tubes, and so are less likely to cause serious damage.

HORIZONTAL TUBE TYPE

These consist of horizontal tubes expanded into vertical tube plates. In this type the air flows through the tubes, while the gases pass over the outside of the tubes.

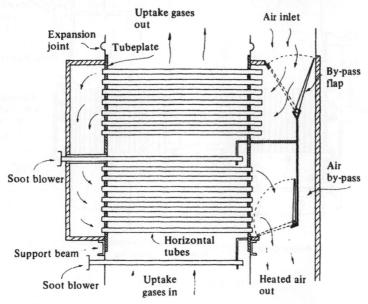

Fig. 53   Tubular type air heater

This form of tube arrangement is less susceptible to chokage, and is more effectively cleaned by soot blowers. But again if allowed to operate at unduly low temperature conditions, heavy deposits, which are often hard enough to resist the action of the soot blowers, can form.

Once fouled, horizontal tube type air heaters are prone to uptake fires, which can rapidly spread throughout the heater and result in serious damage to both the uptake and the air heater.

Horizontal tube heaters are more difficult to clean manually than the vertical tube type, and usually water washing is necessary in order to remove all the deposits.

Again at low load conditions the air heater should be by-passed. This is normally only possible on the gas side. In some cases this can lead to unduly high metal surface temperatures giving greater risk of fire, and in some boilers both the gas and air sides can be by-passed to avoid this possibility.

Q.   Sketch and describe a regenerative type air heater. State where it would be fitted, and give suitable materials for the heat exchange matrix.

A.   In many boilers, especially those of radiant heat type with extensive economiser surfaces, the emerging gas temperature is too low to permit the use of tubular type air heaters, but if desired a regenerative rotary type can be fitted as a last stage heat exchanger.

65

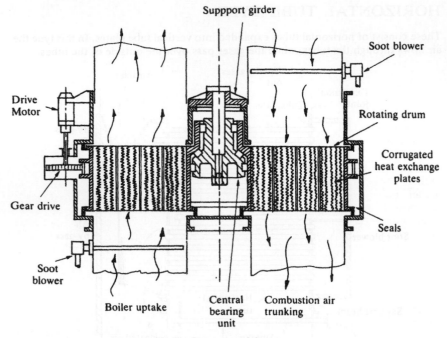

Fig. 54   Rotary type air heater

As shown in Fig. 54, a rotating drum contains a heat exchange matrix of corrugated steel plates so arranged as to pass alternately through the separate gas and air ducts as the drum rotates. The plates absorb heat from the hot flue gases and then reject it to the incoming combustion air.

A motor drives the drum at about 2–4 rpm. A stand-by motor is fitted, but in the event of complete drive failure, by-passes are arranged for both gas and air flows to avoid overheating the stationary drum, or for use at low load conditions to prevent undue fouling resulting from the low operating gas temperatures.

The corrugated plates are made of vitreous-enamel-coated mild steel, or from 'Corten' steel which offers good resistance to corrosion under these conditions. The plate stacks can be removed through an access door for cleaning or renewal.

The drum is carried by a central support bearing, with support jacks provided for overhaul. Mechanical seals are fitted to prevent gas or air being carried over between the ducts, except for the small amount trapped between the plates. It should be noted that when required for an inert gas system on a tanker, the gases are drawn off before the air heater.

Soot blowers are fitted for cleaning the unit, and must be operated for at least one complete revolution of the drum. These blowers are normally operated more frequently that those on the rest of the boiler, due to the greater amount of fouling taking place as a result of the relatively low temperatures at which the unit operates.

In some cases a separate steam air heater is fitted to preheat the air entering the rotary heater, or to heat the air being by-passed.

CHAPTER 5

Boiler Mountings

Q. Enumerate the boiler mountings fitted to a water tube boiler. State the purpose of each fitting and any special design features involved.

A. Various valves and other fittings are required for the proper working of the boiler. Those attached directly to the pressure parts of the boiler are referred to as boiler mountings. In general all these mountings must be carefully designed to perform their function correctly. They must be positioned so as to be readily accessible both for maintenance and operation, the later being performed either directly, or indirectly, by means of extended rods, spindles, etc.

For a water tube boiler the mountings usually consist of the following items:

Safety Valves

These are fitted to protect the boiler from the effects of overpressure. The DOT demand that at least two safety valves are fitted to each boiler, but in practice it is usual to fit three safety valves—two on the steam drum, and one on the superheater outlet header. This latter valve must be set to lift before the drum safety valves so as to ensure a flow of steam through the superheater under blow off conditions. it is normally of the same basic type fitted on the drum.

Main Stop Valve

This is mounted on the superheater outlet header, and enables the boiler to be isolated from the steam line. If two or more boilers are fitted supplying steam to a common line, the stop valve on each boiler must be a screw down, non-return type. This is to prevent steam from the other boilers flowing into a damaged boiler in the event of a loss of pressure due to a burst tube. In some cases the main stop valve incorporates an automatic closing device, designed to operate in emergency conditions, to shut off steam from the main turbines.

Auxiliary Stop Valves

This is basically a smaller version of the main stop, fitted for the purpose of isolating the boiler from the auxiliary steam lines. Again these must be screw down, non-return type valves if necessary to prevent steam flowing back into the boiler in the event of damage. The valve will be mounted on the superheater outlet header; a de-superheater can be used to reduce the steam temperature as required.

Feed Check Valves

These are fitted to give final control over the entry of feed water into the boiler. They must be screw down, non-return valves so that, in the event of a loss of feed pressure, the boiler water cannot blow back into the feed line.

Main and auxiliary feed checks are fitted. The main check is often fitted to the economiser inlet header; if not, like the auxiliary check, it will be mounted directly on the steam drum.

Extended spindles are usually fitted so the checks can be operated from a convenient position. Care must be taken to ensure the valve can be operated easily and quickly, and that a positive indication of the open and closed positions for the valve is given.

Boiler Feed Water Regulator

The water level in a boiler is critical. If it is too low, damage may result from overheating; too high and priming can occur with resultant carry-over of water and dissolved solids into superheaters, steam lines, etc.

Automatic feed regulators are therefore fitted to control the flow of water into the boiler and maintain the water level at its desired value.

They are fitted in the feed line, before the main feed check. In most cases they use a float or thermal means of operation and thus must have a direct connection to the steam and water spaces as required. The regulator can be attached directly to the boiler shell, or alternatively mounted in an external chamber with balance connections to the steam drum, or boiler shell.

In the case of water tube boilers with their high evaporation rate and small reserve of water the control of the water level is so critical that the classification societies demand that some form of automatic feed regulator must be fitted.

Water Level Indicators

The DOT demand that at least two water level indicators must be fitted to each boiler. In practice the usual arrangement consists of two direct reading water level gauges mounted on the steam drum, and a remote reading indicator placed at a convenient control position.

Low Water Alarms

The classification societies demand that these should be fitted to reduce the risk of damage in the event of a loss of water in the boiler due to a burst tube or failure of the feed supply.

In some cases they are mounted inside the steam drum, but many are mounted externally. Various types are available, either steam or electrically operated. Some versions also incorporate high water level alarms.

Blow Down Valves

These are fitted to the water drum to enable water to be blown from the boiler in order to reduce the density. When the boiler is shut down these valves can be used to drain it. They usually consist of two valves mounted in series, arranged so that the first valve must be full open before the second can be cracked open; i.e. sufficient to give the required rate of blow down. In this way the seating of the first valve is protected from damage, so reducing the risk of leakage when the blow down valves are closed.

These blow down valves discharge into a blow down line leading to a ship-side discharge valve.

Scum Valves

These should be fitted when there is a possibility of oil contamination of the boiler. They are mounted on the steam drum, having an internal fitting in the form of a shallow pan situated just below the normal water level, with which to remove oil or scum from the surface of the water in the drum. These valves discharge into the blow down line.

Drain Valves

These are fitted to headers, etc., so enabling the boiler to be completely drained. They must not be used to blow down when the boiler is steaming under load, only being opened when the boiler is shut down or in an off load condition.

Air Vents

These are fitted to the upper parts of the boiler as required to release air from drums and headers, either when filling the boiler, or raising steam.

Superheater Circulating Valves

These are fitted so that when raising steam they can first release air from the superheater, and then provide enough circulation to prevent overheating by allowing sufficient steam to blow off to the atmosphere or to a suitable drain system. They should only be closed when there is enough demand for superheated steam to provide the minimum circulation of steam required to prevent overheating.

Chemical Dosing Valves

These are fitted to the steam drum to enable feed treatment chemicals to be injected directly into the boiler.

Salinometer Valves

These are fitted to the water drum to enable samples of boiler water to be drawn off so that the tests required for the control of the feed treatment can be carried out. At high pressures it is necessary to provide some means of preventing flash off taking place as the pressure over the sample is reduced to atmospheric. This is usually done by passing the water from the salinometer valve through a cooling coil which reduces its temperature to a value below 100°C.

69

Soot Blower Master Steam Valves

These are usually mounted on the superheater outlet header to ensure the superheater is not starved of steam while blowing tubes.

In some cases two valves are fitted in series, with a drain valve between them, in order to prevent steam leaking into the soot blower steam supply lines when these are not in use.

Pressure Gauge Connections

Fitted as required to steam drum, superheater outlet header, etc. to provide the necessary pressure readings. In addition suitable connections must be provided for the pressure sensing points required for automatic combustion control systems, etc.

Thermometers

Pockets must be provided in superheater headers, etc. for the fitting of either direct or remote reading thermometers.

Q. Sketch and describe a water level indicator suitable for boilers operating at low pressures.

A. Every boiler must have at least two independent means of indicating the water level. These indicators must be placed in a position where they can be easily and clearly seen by the operator. Scotch boilers must have the two indicators placed one to each side.

Some small, vertical low-pressure boilers may have three test cocks, placed vertically one above the other in way of the normal water level, together with one gauge glass. However no boiler whose working pressure is above 820 kN/m², or with a diameter in excess of 1·8 m, is allowed to use test cocks as a means of indicating water level.

In most cases marine boilers make use of two indicators in which the water level is clearly visible. For pressures up to 1750 kN/m² these usually take the form of tubular-type water level gauges, while for pressures above this value some form of plate-type gauge glass is used.

The general arrangement of a tubular gauge glass is shown in Fig. 55. It consists of two gun-metal bodies, the glass tube fitted between them, being held in place and sealed at each end by nuts tightened onto soft, tapered sealing rings.

Isolating cocks are fitted to the steam and water connections, while a drain cock fitted at the lower end allows for blowing through the glass to test it. In some cases, gun-metal cocks having tapered plugs stemmed with asbestos are used. These are not only difficult to keep tight, but also are liable to seize up, and in most cases the more efficient asbestos sleeve-type cocks are used.

The gauge may be fitted directly to the boiler shell, and while this is recommended, in some cases—especially with Scotch boilers—in order to clear the smoke box, etc. it is fitted to an external pipe, or alternatively to a hollow or solid column, the ends of which are then connected to the boiler shell by pipes. Isolating cocks or valves, fitted where these pipes are attached to the shell, give the advantage of a double shut off. This enables maintenance to be carried out on the gauge glass cocks while the boiler is still steaming.

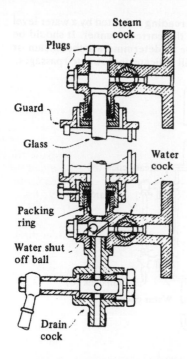

Steam cock

Plugs

Guard

Glass

Water cock

Packing ring

Water shut off ball

Drain cock

Fig. 55   Tubular type water level gauge

A ball valve is fitted to the lower end of the gauge in order to shut off the water in the event of the glass fracturing. Otherwise, as water is blown out, the reduction in pressure causes it to flash off into large volumes of scalding steam, with possible injury to boiler-room personnel. Although steam will escape at the other end, it will not be subject to this great increase in volume, and thus the mass of steam blowing out is limited. However, in some cases restricting orifices are fitted in the steam end, only allowing a greatly reduced quantity of steam to blow out in the event of failure—just sufficient to indicate fracture.

Plugs are fitted in the gun-metal bodies to allow for the renewal of the glass tube, and for cleaning the various passages.

Plate-glass guards should be fitted to prevent injury in the event of the glass tube shattering, especially when blowing through.

Difficulty is often experienced with this type of glass in ascertaining whether it shows completely empty or completely full of water. It is good practice to place a board, painted with diagonal black and white stripes, behind the glass. If it is full of water, refraction will cause the stripes to appear bent to the opposite angle. If no board is fitted, a pencil, or similar object can be used to the same effect.

Finally the handles of the steam and water cocks must lie vertically downwards when in the full open position, while the drain cock handle must be in this position when closed. This is to prevent vibration, etc. causing the cocks to move to such a position that a false indication of the water level could result.

Q.   Describe how you would test a water level gauge of tubular type to ensure it is indicating a correct level. State the various causes for a false reading.

71

A. If any doubt exists about the accuracy of the reading indicated by a water level gauge it should be tested by blowing through in the correct manner. It should be noted that, when testing, it is not usually possible to determine whether steam or water is issuing from the drain, but a strong blow will indicate unobstructed passages.

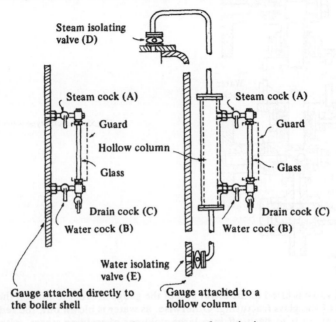

Fig. 56   Alternative arrangements of water level gauges

Fig. 56 shows two common means of attaching water level gauges to the boiler. The following test procedure should be carried out on gauge glasses attached directly to the boiler shell:

Close the water cock (B), and open drain cock (C). A strong blow will indicate the steam cock (A) is clear. Then closing cock (A), open cock (B); a strong blow now indicates the cock (B) is clear.

A similar procedure can be carried out for gauges attached to the boiler by means of an external pipe, or by means of a solid column. In this latter case, the isolating cocks on the pipes leading to the column may also be choked. The advantage of the hollow column-type fitting shown in Fig. 56, is that in the event of chokage it is possible by blowing to determine which of the four cocks is choked.

To do this first carry out the test procedure previously described using cocks (A) and (B) together with drain cock (C). In both cases a strong blow through the drain indicates these cocks are clear. Then to check the isolating cocks (D) and (E) together with the hollow column itself, use a procedure known as cross blowing. For this close cocks (A) and (E), a strong blow through the drain now indicates cocks (D) and (B) and the respective passages are clear. Then close (D) and (B) and open cocks (A) and (E), where again a strong blow shows these cocks and passages to be clear. If however during these operations only a weak blow, or a no blow occurs, it indicates an obstruction is present in whichever passage is open at the time, and a simple process of elimination will show where the fault lies.

A problem can arise with the hollow column-type fitting in that, if either of the isolating cocks (D) or (E) fitted on the boiler chokes, a strong blow will still be obtained if the gauge glass is tested by means of cocks (A) and (B) only, and no indication of the obstruction will be given. Thus when the gauge glass is returned to service, if say cock (D) on the boiler is choked, there will be a tendency for the water level in the glass to rise above that in the boiler, due to the vacuum forming in the steam balance connection to the boiler, causing water to flow into the hollow column. If cock (E) is choked, the water again rises in the glass due to condensation, until such time as testing is carried out. Then although a strong blow is obtained, no water will appear in the glass when it is returned to service. Thus the full test procedure, described previously for this type of fitting, should be carried out if any doubt arises about the accuracy of the level indicated.

Having checked that all cocks and passages are clear, and with the steam and water cocks closed, and drain open, a tubular-type water level gauge should be put back into service as follows:

Close the drain, and open water cock (B) very slowly. If the water level in the boiler is above the passage to this cock, it will enter the glass and rise slowly to the top. Now open the steam cock (A), and the correct water level in the boiler will be indicated; this should be between one quarter to three quarters of the glass.

If the level drops from the glass when the steam cock is opened, it indicates the water level in the boiler lies somewhere between the water passage to the gauge and the bottom of the glass. In a Scotch boiler the bottom passage lies at least 100 mm above the top of the combustion chamber; the boiler is still in a safe condition, but feed water should be added, or the supply increased immediately. On the other hand if the water remains at the top of the glass, it indicates too much water in the boiler, and the feed supply should be reduced, or stopped if necessary, to prevent priming.

If no water appears in the glass when the water cock (B) is opened, the level of water in the boiler must be below the gauge water passage and the boiler is in a potentially dangerous condition. Reduce boiler load and rate of firing and, as long as the water covers the combustion chamber tops, the feed supply should be increased, and if necessary the stand-by feed pump and auxiliary feed check put into operation. But if the water level does not reappear quickly, or if any doubt exists about the water covering the combustion chamber tops, take the boiler out of operation. Shut off the fuel and, if it is suspected that overheating has occurred, operate the easing gear on the safety valves to release the boiler pressure.

When the boiler has cooled down it should be examined, and only returned to service when considered to be in a safe condition.

Difficulty in deciding whether the glass is completely empty or completely full of water, can be avoided by making use of the principles of refraction. A pencil, or similar object, held at an angle behind the glass tube will appear to be bent to the opposite angle if the tube is full of water.

A plate-glass guard should be in place around the glass tube to reduce the risk of injury during these test procedures.

Q. Enumerate the faults to which direct reading gauge glasses of tubular type are subject. Explain the effect these faults will have upon the level indicated in the glass.

A. Various reasons can lead to a false reading being indicated by a tubular water level gauge glass. Choked cocks and passages are a common one.

If the steam cock is choked, a vacuum forms in the upper part of the glass causing the water to rise until it completely fills the glass.

If the water cock is choked, the water level will rise slowly, condensation in the steam space gradually filling the glass

Similar results, although to a lesser extent, can be caused by partially choked cocks and passages. This chokage is generally brought about by a build-up of deposits left behind by boiler water evaporating away due to leakage in way of cocks, etc.

Another reason for obstruction is the cock handle being twisted; so causing the plug to be closed, although the handle is in the open position. This can occur due to tapered plugs, stemmed with asbestos, being nipped up too tight in an attempt to stop them leaking. This type of fault is unlikely with asbestos sleeve-type cocks, provided reasonable care is taken when inserting a new sleeve. A locating rib on the side of the sleeve ensures correct alignment. Also ensure the cock handle is correctly attached to the spindle.

It should be noted that the handles on the steam and water cocks should always point vertically downwards when in the open position as another fault can arise if they are left pointing upwards; i.e. vibration causing a loose plug to move round to a closed or partially closed position.

The glass tube itself can cause trouble. If it is too short it may result in the packing sleeve being squeezed over the ends of the glass, so blocking the opening. This is most likely to happen at the top steam end of the glass. Again too long a glass projecting up into the top cock body can obstruct the steam passage.

A dirty or salted glass can prevent the water level being easily seen.

Where water level gauges are fitted to the boiler indirectly by means of a solid or hollow column, as shown in Fig. 56, the isolating cocks on the boiler can become choked. This again causes the water level in the glass to read too high—for similar reasons to those stated for the steam and water cocks. Additional care must be taken with hollow column-type mountings in that blowing through by means of the steam and water cocks on the gauge itself, and not using the isolating cocks on the boiler, will not indicate if either of the latter is choked.

Gauges fitted to the boiler by either of the above methods should be checked on a new boiler, or after a refit, to ensure that no distortion has taken place or bends formed in which condensation can collect; these may cause a possible false reading.

It should also be noted that the water cools down in the relatively long pipes connecting the columns to the boiler; this changes its density, resulting in a level in the glass slightly below that in the boiler. Immediately after blowing, the level in the glass will appear slightly higher—now corresponding to that in the boiler—until the water again cools down.

The water level gauge should always be kept clean, well lit, and in good order. All passages and cocks liable to chokage should be inspected, and cleaned if necessary when the boiler is shut down.

The water level gauges should be tested at frequent intervals, and always if any doubt arises about the accuracy of the water level indicated. Any obstruction must be cleared as soon as possible.

Steam and water cocks and passages in the gauge can be cleared while the boiler is still steaming. To do this, shut the steam and water cocks, and open the drain. Remove the cleaning plug opposite the obstruction. If this still cannot be cleared, screw in a plug with a small hole, about 5 mm diameter, drilled through it in place of the cleaning plug. Insert into this hole a rod of such a size that, held by a gloved hand, it can be moved easily without being slack. Then open the choked cock and push the rod through to clear the blockage. When clear, the open drain will prevent a build-up of pressure, and only a small amount of steam will blow past the rod, the glove protecting the operator from injury. Then close the cock and replace the normal cleaning plug. The gauge glass can now be tested and, if satisfactory, returned to service.

Do not carry out this operation on a plate-type glass on a high pressure boiler.

Q. Sketch and describe a water level indicator suitable for boilers working in the medium pressure range.

A. Although plate-type water level gauges can be used for low pressures, in view of their greater cost, they usually only come into general use for pressures above a value of about $1750\,kN/m^2$. The reflex glass shown in Fig. 57 is an example of a type of water level indicator suitable for boilers working at a medium pressure range between about $1700\text{--}3000\,kN/m^2$.

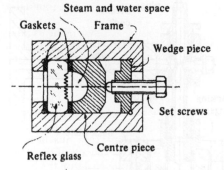

Fig. 57 Reflex type water level gauge

These gauges are supplied with gun-metal bodies up to pressures of $1750\,kN/m^2$, while for higher pressures forged steel bodies are used.

The gauge is normally fitted directly to the boiler shell or steam drum. The isolating cocks in the steam and water connections, together with the drain cock, are of the asbestos sleeve type. In many cases these cocks can be operated by extended rods, or chains, to prevent injury in the event of the glass shattering when blowing through the glass. However, the glass plates are so strong that this form of failure does not occur often, and even then the pieces of glass frequently remain in position; thus the fitting of an external guard is not necessary.

The single-sided glass ingeniously makes use of the refraction of light so that, when illuminated from the front, the series of ribs at the back of the glass plate cause the light rays to be reflected back from the steam space and absorbed in the water space. This gives a bright silvery appearance to the former, while the latter shows dark. The strong contrast between the two enables the operator to see immediately the position of the water level. It also makes it possible to tell at a glance whether the glass shows completely empty or completely full of water—a condition which causes some confusion with many other types of gauge glass.

A ball valve is normally fitted to the lower end of the gauge to shut off the water in the event of glass plate shattering and blowing out. The ball valve thus prevents the escape of water, with the resulting flash-off of large amounts of steam, making it difficult to shut off the gauge, and possibly causing injury.

75

Another problem arises at higher pressures due to the fact that hot distilled water at high pressure erodes the glass away. This effect is even more pronounced if alkaline feed additives are being used. To resist this a special borosilicate glass is used for the higher pressures, but can only give limited protection. For pressures above 3400 kN/m² some means must be used to prevent the water from coming into direct contact with the surface of the glass. This is usually done by placing a sheet of mica between them. Due to the ribbed glass, the reflex-type gauge cannot make use of this form of protection, and so is not suitable for use with high pressure boilers.

Q. Sketch and describe a type of water level indicator suitable for use on high pressure water tube boilers.

A. All boilers must have at least two independent means of indicating the water level, and in the case of high pressure water tube boilers working at values above

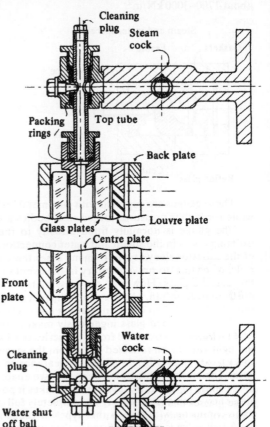

Fig. 58   Double plate type water level gauge

76

3400 kN/m^2, these usually take the form of double-sided, plate-glass type water level indicators, with mica protection for the glass plates. This protection is necesssary as at high pressures hot distilled water erodes the glass away, and unless a sheet of mica is placed between the glass and the water, attack take place quickly, indeed at the higher pressure ranges the glass will burst within a few hours if this protection is omitted.

The general arrangement of a typical double-plate water level gauge is shown in Fig. 58. It consists basically of a hollow centre piece with flats machined on each side clamp plate. Care must be taken during assembly to prevent undue stresses being set up which will cause the glass plate to shatter when put into service. Thus the following procedure should be carried out. Strip down the faulty gauge. Discard the used glass plates, mica sheets, and joints. Make sure all joint faces are scrupulously clean. Check frame and cover plates for flatness; any warping can cause the glass to shatter. Build up the gauge, inserting the new joints, together with the mica sheets, in their correct sequence. This is indicated in Fig. 59. Bolt the clamp plate onto the outer case. The clamping bolts should be pulled finger tight onto the louvre plate. Then starting from the centre, tighten these nuts in the order indicated in Fig. 60. Do not overtighten, and pull up evenly, preferably using a torque spanner.

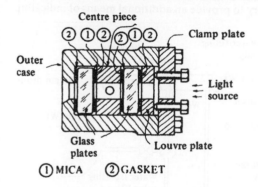

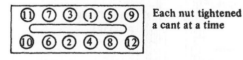

Fig. 59   Plan view of gauge

Each nut tightened
a cant at a time

Fig. 60

The louvre plate at the back of the gauge is placed with its slots angled upwards so that it directs the light rays from the electric lamps in such a manner that the actual water level in the glass appears as a plane of light when viewed from below.

A ball valve is fitted at the lower end of the gauge to shut off the water in the event of the glass plates shattering. It should be noted that some forms of double plate gauge glasses can be placed on the boiler upside down. This places the ball valve at the top of the gauge, where it rolls down and obstructs the steam passage, so causing a false reading. It is thus advisable to mark this type of fitting so as to clearly identify the top and bottom ends of the gauge.

When installing a new gauge glass, first shut the steam and water cocks, and open the drain. Remove the defective unit and fit the new gauge. Leave it in this condition,

77

with the steam and water cocks closed and drain open, to heat up for some hours. Then just crack open the steam cock. After about twenty minutes follow up the clamp nuts in the correct sequence, preferably using a torque spanner. Then close the drain, and fully open the steam and water cocks, to put the gauge into operation. Do not stand directly in front of the gauge during these operations in case the glass shatters.

Strip down and rebuild the defective unit as soon as possible and place it in a rack close to the boiler.

The shut off and drain cocks are asbestos sleeve type, and can be operated at some safe distance from the gauge.

Two of these gauges are normally fitted directly onto each boiler steam drum, usually to one of the end plates, but in some cases to the side of the drum—one at each end.

Q. Sketch and describe a remote reading type water level indicator suitable for a high pressure water tube boiler.

A. Difficulty is often experienced in observing the water level as indicated by the direct reading water level gauges mounted on the steam drums of water tube boilers. Thus it is usually considered necessary to provide an additional means of indicating

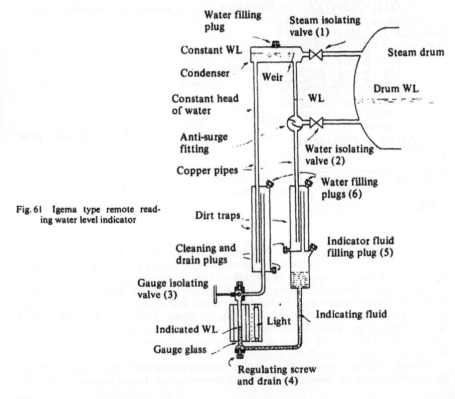

Fig. 61 Igema type remote reading water level indicator

the water level at some point convenient to the starting platform or control room. This can be done by a remote reading indicator such as the (Igema) gauge shown in Fig. 61.

This consists basically of a U-tube, the two legs connected to the steam drum as shown. Red indicating fluid, which is insoluble in water, fills the lower end and remains there since its density is greater than that of the water.

Above this fluid the two legs of the U-tube are filled with water; one being kept filled to a constant head by means of steam condensing in the unlagged condenser. The level in the other leg corresponds to that in the steam drum. Thus the heads supported by the indicating fluid vary. As the water level in the drum rises so it tends to balance the constant head, and the indicating fluid rises in the glass. The opposite happens when the drum water level falls, the level of the indicating fluid in the glass also falling.

The sharp contrast between the red indicating fluid and the water enables the operator to see the indicated water level at a glance. A completely empty or full glass is immediately obvious.

When taking the boiler out of service, shut off the remote indicator by first closing the gauge isolating valve (3), then the steam isolating valve (1), and finally the water isolating valve (2).

When opening up, first open the steam valve (1), then the water valve (2), and finally the gauge isolating valve (3).

If the remote indicator is connected to the same balance connections as one of the direct reading water level gauges, it is important that the remote indicator is isolated before the water level gauge is blown through. Otherwise water may be drawn out of the legs of the U-tube so causing a false water level to be indicated by the remote reading gauge.

After cleaning, etc. the following procedure should be carried out to refill the indicator. First close the isolating valves (1) and (2) on the boiler, and the regulating screw (4). Remove all filling plugs. Then pour in the indicating fluid through the indicator filling plug (5) until the lower part of the U-tube is completely filled, fluid overflowing at the filling plug.

Close the gauge isolating valve (3) and replace the filling plug (5). Then slowly pour distilled water into the water filling plugs (6) on top of the dirt traps until it overflows. Replace the filling plugs. Finally pour water into the top filling plug (7), until again it overflows; the plug is then replaced.

The remote reading gauge glass should now show completely red. Leave it in this condition until full boiler pressure has been raised.

When the boiler is under steam open the steam valve (1), then the water valve (2) followed by the gauge isolating valve (3). Leave for about 15 minutes to settle down, then crack open the regulating screw (4) and slowly bleed off excess indicating fluid, dropping the level about 6 mm at a time, with about fifteen minutes between, until finally the level of indicating fluid at the centre of the glass corresponds to the water level at the centre of the direct reading water level gauge glass.

Although subjected to boiler pressure, the remote indicator glass is not at high temperature and very rarely gives trouble. However, the apparatus should be cleaned out about once every six months. The indicator should be isolated, drained, and the flushed through with clean water. The indicator must never under any circumstances be blown through either with steam or water.

The glass is illuminated from behind, access to these lights being obtained by removing the sheet metal casing at the back of the gauge.

Q. Sketch and describe a low water alarm suitable for a water tube boiler. State the means of operation, and why these alarms must be fitted to water tube boilers.

A. Unlike the procedure with Scotch boilers, if the water level in a water tube boiler disappears from the bottom of the water level gauge glass, immediate action must be taken, the boiler being put out of service.

Due to the high evaporation rate and small reserve of water in a water tube boiler, for various reasons such as a malfunction of the feed water regulator, or feed pump, or by a burst tube within the boiler itself, this loss of water can happen very rapidly. The classification societies therefore demand the fitting of a low water protection device. This must both sound an alarm and cut off the fuel oil to the boiler.

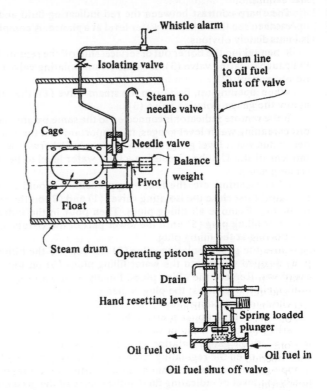

Fig. 62   Low water alarm

Many different types may be fitted; a typical steam operated low water alarm and fuel oil cut off is shown in diagrammatic form in Fig. 62.

The device consists basically of an operating mechanism in the form of a float controlled needle valve, an alarm whistle, and a fuel cut off valve. When the water level falls below a certain predetermined value the float drops—so opening the needle valve. This allows steam to pass to the whistle alarm and to the actuator of the fuel oil shut off valve.

When the steam pressure acts on the actuator piston it pushes it downwards, so closing the shut off valve. When the valve closes, a spring loaded plunger moves into a slot in the valve spindle, and holds the valve in the closed position until it can be reset by hand. This is to prevent some transient event, such as the water surging in the drum due to heavy weather conditions, causing the fuel oil shut off valve to first close and extinguish the furnace flame. If this valve were then allowed to reopen automatically immediately the water level is re-established, it could result in large quantities of unburnt oil being sprayed into the hot furnace leading to a possible gas side explosion causing damage or starting an uptake fire.

In addition to this resetting device, an anti-surge device will be fitted to the operating mechanism. In the type being considered it takes the form of a cage placed around the float. Holes in this cage are such that if, during heavy weather, the water level drops momentarily below the minimum level allowed, it will have surged back before the water in the cage has time to drain out enough to operate the mechanism.

Another refinement which may be fitted is the internal steam supply pipe. This prevents deposits forming around the needle valve seat in the event of leakage, i.e. as might occur if the chamber were left open to the boiler water.

In many versions the operating mechanism is mounted in an external chamber, and connected to the drum by steam and water balance connections. This gives the advantage that maintenance can be carried out on the operating mechanism without shutting down the boiler.

The low water alarm should be tested at frequent intervals, and as soon as practicable after any work or adjustments have been carried out on it. If for any reason the device has to be shut off a notice should be displayed on the starting platform, or in the control room, informing the watch-keeping Engineers of the fact. The device should be returned to service as soon as possible.

Q. Sketch and describe an electrically operated low/high water level alarm and cut-out.

A. The type illustrated in Fig. 63 operates on the principle that while unlike magnetic poles attract each other, like poles are repulsed.

An external float chamber with balance connections to the steam drum is mounted in way of the drum water level. A float within the chamber then relates to this water level, and moves an attached magnet up and down in response to any changes of water level within the steam drum.

Four micro switches with magnetically operated arms are fitted on the outside of the balance chamber. They are arranged so that, when the internal magnet comes level with one of these external magnets, like poles are presented, and so, repelling each other, operate the micro switch.

The positions of the four switches are set to correspond to predetermined water levels within the steam drum. One switch will correspond to a low water level condition, and when this is reached will operate an alarm. If then no corrective action is taken, or it proves ineffective, and water level continues to fall, the next switch positioned to a predetermined low low water level will trip out the fuel oil shut-off valve, so shutting the boiler down before damage can occur.

The other two switches are positioned in a similar manner for high water levels, operating an alarm for a predetermined high water level in the drum, tripping out the fuel oil shut-off valve, and closing in a motorised feed check valve in the event of a high high water level.

81

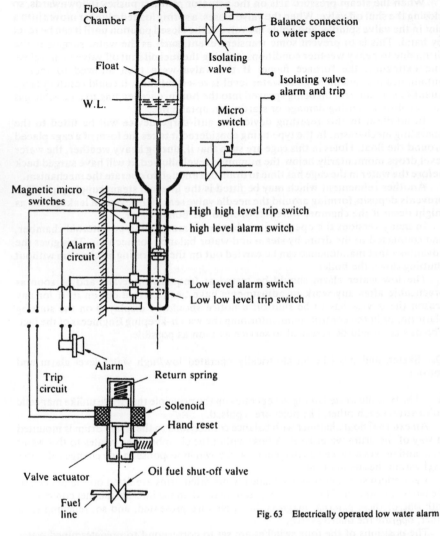

Fig. 63 Electrically operated low water alarm

Restrictions in the balance connections, or time delay switches in the electric circuit, prevent operation of this protective device by some transient event, such as surging of the drum water level due to movement of the ship in a heavy seaway.

Micro switches are also fitted to the isolating valves on the balance connections to ensure that they are fully open when the device is in operation.

As a further safeguard it is usual to fit two different types of protective device for this duty, and the unit described could well be used in conjunction with a differential pressure transmitter which compares the variable drum water level with a constant head of water provided by a condensing reservoir connected to the drum steam space.

Movement of the transmitter is amplified and used to operate four micro switches in a similar manner to the magnetic type, as the water level moves through its full range.

These devices should be tested by actually raising or lowering the drum water level when the boiler is off load, the direct-reading water level gauges being observed throughout the test to ensure correct operation of the devices. Each device is gagged in turn, to make sure that each individual unit is functioning properly. It should be noted that differential pressure devices are sensitive to water density and should only be tested with the boiler at full working pressure.

Q. Sketch and describe an Improved high lift type safety valve.

A. This is a safety device fitted to the boiler to prevent overpressure. The DOT demand a minimum of two safety valves to each boiler, and although these may be fitted in the same valve chest with a single connection to the boiler, the safety valves must be capable of releasing the maximum amount of steam the boiler can evaporate, while still keeping within the 10 per cent accumulation of pressure rule.

This rule is necessary because, having calculated the cross sectional area for the valve bore, the valve lid must be able to lift at least $\frac{1}{4}$ of the valve bore in order to provide full steam flow. However as the valve lifts, the force to compress the spring also increases, and so the higher the valve lifts the greater the increase in boiler pressure. The DOT limit this accumulation of pressure to 10 per cent of the maximum allowable working pressure for the boiler. This means that the lift of an ordinary spring-loaded mitre valve, although mechanically able to lift $\frac{1}{4}$D, would be very limited under these conditions, and the valve would have to be very large to give the required throughput.

Various designs have been developed to increase this working lift, and the Improved high lift safety valve is a very suitable one for low and medium pressure boilers. See Fig. 64.

These valves increase their lift over that of a simple spring-loaded mitre value in two ways: See Fig. 65.

One is the specially shaped valve seat, the other being to use the lower spring carrier in the fashion of a piston, which acted upon by the pressure of the waste steam helps to compress the spring.

As can be seen in Fig. 65, a lip is placed around the valve seat so that, when the valve lid lifts, escaping steam is trapped in the annular space around the valve face; the resultant build-up of pressure acting upon the greater valve lid area causes the valve to lift sharply. This arrangement gives another advantage in that if the valve were allowed to close slowly, the final film of steam blowing across the faces sets up high frictional resistance to the closure of the valve, and this can cause the boiler pressure to fall well below the blow off pressure before the valve finally closes. This is referred to as the blow down effect, and can cause considerable wastage of steam. The shaped valve seat causes the valve lid to pulsate, and this 'Dirling' action enables the valve to close cleanly and sharply with very little blow down effect.

The second feature of the valve lies in the fact that as the escaping steam enters the waste pipe and, although regulations exist to prevent this pipe being made too small, some build-up of pressure does take place. The Improved high lift safety valve makes use of this waste pressure to increase the valve lift. This is done as previously mentioned by allowing this pressure to act upon the lower spring carrier, which fits

83

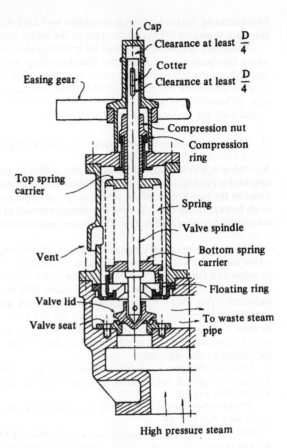

Cap

Clearance at least $\frac{D}{4}$

Cotter

Clearance at least $\frac{D}{4}$

Easing gear

Compression nut

Compression ring

Top spring carrier

Spring

Valve spindle

Bottom spring carrier

Vent

Floating ring

Valve lid

Valve seat

To waste steam pipe

Fig. 64 Improved high lift safety valve

High pressure steam

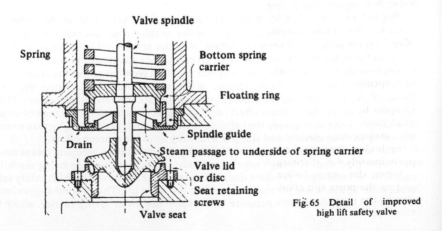

Valve spindle

Spring

Bottom spring carrier

Floating ring

Spindle guide

Drain

Steam passage to underside of spring carrier

Valve lid or disc

Seat retaining screws

Valve seat

Fig. 65 Detail of improved high lift safety valve

within a floating ring, so forming in effect a piston; the pressure acts upon this piston causing it to move up, helping to compress the spring, and so increasing the valve lift.

One of the features of this design is the loose ring, which is arranged so that if the spring carrier were to seize in the ring, the ring itself would lift and the overall action remain unchanged.

The valve lid is wingless, and has a flat faced landing. Thus it must be ground in by means of a jig.

Adjustment of the valve is carried out by means of a compression nut screwing down onto the top spring plate. A compression ring is fitted after final adjustment to ensure no further movement takes place. A cap is then placed over this compression nut and the top of the valve spindle, and a cotter is passed through and padlocked to prevent tampering by unauthorized persons. Clearances between this cap, the valve spindle and the cotter are such as to prevent the valve being held down externally, whether by accident or design.

A hexagon on top of the cap enables it to be turned, this motion being transmitted through the cotter and the valve securing pin to the valve lid, turning it upon the seat. With saturated steam this action is often effective in stopping the valve feathering and allowing it to seat firmly.

Easing gear is fitted so that in the event of an emergency the valve can be opened by hand to a full lift of $\frac{1}{4}$D to release the boiler pressure.

An open-ended drain must be fitted to the waste pipe to prevent any build-up of water in the pipe causing a head of water to form over the valve lid so increasing the blow off pressure. Also in very cold conditions this water could freeze in the upper parts of the waste pipe with possibly disastrous results.

Q. Sketch and describe a full bore safety valve operated by means of a pilot control valve.

A. Conventional spring-loaded safety valves present problems when called upon to deal with steam at high pressures and temperatures. These include distortion of the spring and valve seats, and the difficulty of getting the valve to close smartly at these extreme conditions. The latter is necessary in order to prevent feathering; this is the name given to a condition where a thin film of steam blows across between the valve faces—the wire-drawing resulting from this quickly cutting the valve faces and leading to loss of steam. This is especially the case when the steam is superheated.

One type of valve designed to deal with these problems is the full bore type safety valve. This consists of main and pilot control valves, both being in direct communication with the boiler pressure. See Fig. 66.

Each safety valve is operated by its own control valve, the latter consisting of a small spring-loaded valve set to operate at the boiler blow off pressure. As it lifts it blanks off ports leading to the atmosphere, and allows steam pressure to build up and act upon the operating piston attached to the main valve spindle. This piston has about twice the area of the main valve, and the forces set up cause the main safety valve to open one-quarter of its diameter, so giving full bore conditions for the escape of the boiler steam to the atmosphere through the waste pipe. The throughput is approximately 4 × discharge capacity of an ordinary safety valve.

When the excess boiler pressure has been relieved the pilot valve closes, so opening the ports and allowing the operating steam to vent to the atmosphere; this releases the pressure acting on the piston. The escaping steam, helped by the valve return spring, closes the main valve.

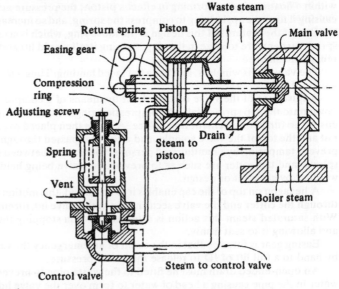

Fig. 66   Full bore safety valve

The main valve is then kept closed by the boiler pressure acting upon it. As there is no spring to oppose this force, the higher the boiler pressure the greater the closing load, this being the opposite case to a direct spring-loaded valve.

When this type of valve is fitted to the superheater, troubles due to the high temperatures involved, can be greatly reduced by mounting the control valve on the steam drum so that it operates with saturated steam. In this way both the control valve spring and the operating piston from the superheated steam are protected.

Other advantages for this type of valve are that the main valve has no heavy spring to be affected by the high temperature steam. The light return spring, being at the other end of the chest away from the escaping steam, is of relatively little importance.

The control valve seat is small and so is less liable to distortion than larger seats and this, coupled with the use of saturated steam, helps to ensure that it closes tightly, thus giving a positive closing action to the main valve. The boiler pressure acting upon this enables it to seat tightly, even against slight distortion of the main valve seat.

Adjustment of the blow off pressure is carried out by changing the compression on the control valve spring; this is done by means of an adjusting nut fitted to the control valve.

Easing gear is fitted to act directly upon the main valve in order to open it in case of emergency. In addition a hand easing lever is also fitted to the control valve, enabling steam to be admitted to the operating piston of the main valve.

An open-ended drain must be fitted to the waste pipe.

Q.   Sketch and describe a full lift safety valve of the nozzle reaction type.

A. A number of direct spring-loaded disc type safety valves obtain a full lift of one-quarter of their diameter within the 10 per cent Accumulation of Pressure rule by allowing the valve disc to lift within a sleeve so that, acted upon by the waste steam pressure, it performs in the manner of a piston helping to compress the spring. Further assistance being provided by reaction forces set up when the escaping steam is deflected by a specially shaped valve seat and disc face.

The Hopkinson Hylif safety valve is an example of this type of safety valve.

When this valve first lifts in response to a condition of overpressure, it allows the escaping steam to act upon the full area of the enlarged face of the valve disc, lifting it up into the fixed guide sleeve. This prevents waste steam pressure acting on top of the valve, and a vent is provided to keep the space between the valve disc and the spindle guide at atmospheric pressure. Thus the waste steam pressure acting upon the face of the valve disc lifts it, so helping to compress the spring.

When the valve fully enters the guide sleeve, it causes some of the escaping steam to be deflected downwards by the bottom edge of the guide sleeve, the resulting reaction set up assisting the waste steam pressure to lift the valve to its full open, $\frac{1}{4}D$ position. It is claimed that this, together with the nozzle shaped inlet to the valve seat, gives full flow conditions for the escaping steam.

When the boiler overpressure has been relieved, the valve begins to close, and as the face of the valve disc emerges from the guide sleeve the reaction effect ceases, and the valve now closes cleanly and sharply. This helps to prevent feathering.

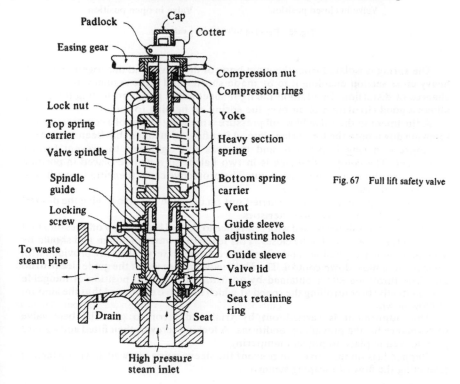

Fig. 67   Full lift safety valve

87

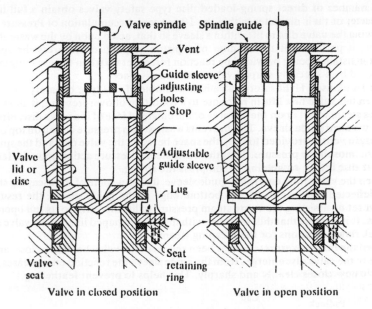

Valve spindle   Spindle guide

Vent

Guide sleeve
adjusting
holes
Stop

Valve
lid or
disc

Adjustable
guide sleeve

Lug

Valve
seat

Seat
retaining
ring

Valve in closed position   Valve in open position

Fig. 68   Detail of full lift safety valve

The spring is isolated from the main body of the valve, and this together with its heavy cross section enabling it to be made shorter and more resilient, reduces the chances of distortion. In addition the light sheet metal guard protecting the spring allows a good circulation of air over the spring, so helping to cool it.

Adjustment to the valve blow off pressure is made by means of a compression nut screwing down onto the top spring plate. After final adjustment has been carried out a compression ring is inserted, and the lock nut tightened to prevent further movement. The compression ring is in two halves so it can be placed in position without disturbing the compression nut. The protective cap is then fitted over the nut and ring, and padlocked in place.

A further adjustment can be carried out on this type of valve to give the desired discharge and blow down characteristics.

The term blow down is used with reference to the fall below blow off pressure that occurs before the valve finally closes. This leads to a waste of steam and should be avoided as much as possible, and many high capacity safety valves are provided with some form of blow down control. The Hylif valve makes use of the guide sleeve to do this. The throttling effect obtained by adjusting the vertical position of the guide sleeve effectively controlling the speed at which the valve closes and thus the amount of blow down.

This adjustment is carried out by trial and error to give the best valve performance for the prevailing conditions. A locking screw is then fitted and a guard plate locked in place to prevent tampering.

Stepped lugs on the valve seat prevent the sleeve being screwed down too far and restricting the flow of escaping steam.

Easing gear is fitted to enable the valve to be opened by hand in case of emergency; this is arranged so that it cannot be used to hold the valve closed. An open-ended drain must be fitted to the waste pipe.

Q. Sketch and describe a consolidated type full lift safety valve.

A. The general arrangement is illustrated in Fig. 69. The valve has a number of special features. First an adjustable blowdown control ring screwed onto the valve seat provides a quick full lift or 'pop' action to the opening valve disc, and a cushioning or blowdown control action as the valve closes. Both actions are provided by deflection of the escaping steam against the inclined sides of the valve disc, so producing an upthrust: with the initial high escape pressure, this gives the valve its opening 'pop' action; then, as boiler pressure falls and the valve closes under the influence of the boost cylinder, this same upthrust, although now reduced, cushions the final seating of valve.

Adjusting the ring too high produces excessive cushioning, so preventing a snap closing of the valve, which can cause excessive loss of boiler pressure before the valve eventually shuts. Set too low, and insufficient upthrust will be provided on the opening valve with resulting loss of the required 'pop' action, which in turn can lead to the valve disc chattering on its seat.

A second feature is a boost cylinder fitted above the valve disc to provide a precision closing control. As the valve pops full open, some escaping steam is bled off to this boost cylinder from whence it is vented to the atmosphere. As boiler pressure falls and the valve begins to shut, this atmospheric vent closes, immediately causing a sharp rise of steam pressure in the boost cylinder. The resulting downthrust on the valve disc produces a positive closing action and gives the desired snap shut which reduces blowdown. The term 'blowdown' refers to the amount the boiler pressure falls below the blow-off pressure before the safety valve finally closes. A snap closing action also prevents feathering of the escaping steam, which can damage valve faces when high superheat temperatures are involved, as a thin film of steam continues to blow across the valve faces.

For high temperature applications a thermodisc valve seat is fitted; this is formed by recessing the bottom of the valve disc to form a thin wall, so providing extra flexibility at the point of contact as the valve closes. This allows better heat transfer between the mating surfaces, which serves to reduce local distortion and so prevent valve leakage. Easing gear is fitted so that in the event of an emergency the valve can be opened by hand in order to quickly release the boiler pressure. This gear must be arranged so that it cannot interfere with the opening of the valve under normal operating conditions. When the safety valve is being overhauled, remember to check that the operating mechanism of the easing gear is in full working order.

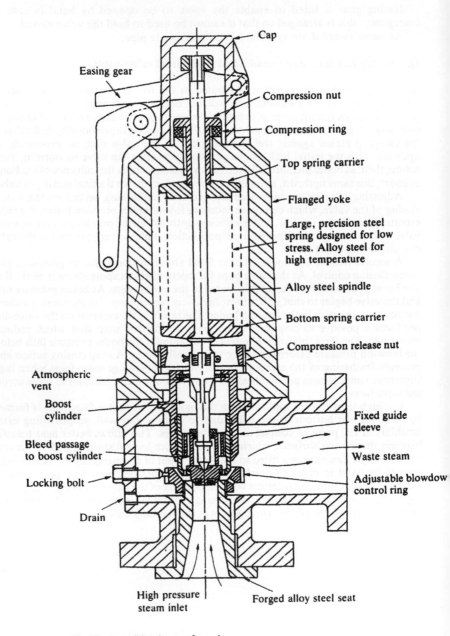

Fig. 69 Consolidated type safety valve

CHAPTER 6

Combustion of Fuel in Boilers

Q. Describe the process by which a residual fuel oil is burnt in a boiler furnace.

A. The combustion of a residual fuel oil in a boiler furnace takes in a number of stages, which are described as follows.

The oil is first heated in steam or electric fuel oil heaters. This reduces its viscosity and makes it easier to pump, filter, and finally to atomize.

However it must not be overheated at this stage, otherwise a process known as 'cracking' occurs, leading to carbon deposits, and the formation of gas in the fuel oil lines, etc. The gas, due to its large volume, reduces the mass of oil passing through the burners, which in turn leads to a possible reduction in the steaming rate of the boiler owing to the reduced amount of fuel actually burnt.

This gasification can also cause instability in the combustion process itself, resulting in a fluctuating flame formation.

The heated oil is now passed through the burners where it is atomized; this process breaks it up into a fine spray of droplets, so presenting a very large surface area of oil to the combustion processes. The droplets formed are of two main types, i.e. very fine particles consisting of the lighter fractions of the fuel, which form a fine mist, and slightly larger droplets formed by the heavier fractions of the residual fuel.

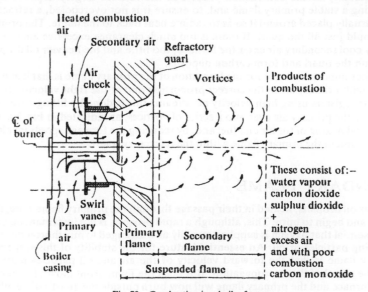

Fig. 70 Combustion in a boiler furnace

The burner also imparts rotational energy to the fuel so that it leaves the burner tip as a hollow, rotating cone formed of fine droplets of oil.

The combustion stage itself can now commence, and in a boiler furnace a type of combustion often referred to as a 'suspended flame' is used. For this a stream of oil particles and air enters the combustion zone at the same rate at which the products of combustion leave it. The actual flame front therefore remains stationary, while the particles pass through it, undergoing the combustion process as they do so. The combustion zone itself can be sub-divided into two main stages; these are referred to as the primary and secondary flames.

PRIMARY FLAME

For the oil to burn, it must be raised to its ignition temperature, where continuous vaporization of the oil required for its combustion takes place. Note this temperature should not be confused with the flash point temperature of the oil, where only the vapour formed above the oil in storage tanks, etc. will burn. The ignition or burning temperature should normally be at least some 20°C above this value.

For the reasons already stated this ignition temperature cannot be obtained in the fuel oil heaters, and therefore the heat radiated from the flame itself is utilized so that, as the cone of atomized oil leaves the burner, the lighter hydrocarbons are rapidly raised to the required temperature by the heat from the furnace flame; they then vaporize and burn to form the primary flame.

The heat from this primary flame is now used to heat the heavier constituents of the fuel to their ignition temperature as they, together with the incoming secondary combustion air, pass through the flame.

The stability of the combustion process in the furnace largely depends upon maintaining a stable primary flame and, to ensure it is not overcooled, a refractory quarl is usually placed around it so as to radiate heat back to the flame. The primary flame should just fill the quarl. If there is too much clearance excessive amounts of relatively cool secondary air enter the furnace; too little and the heavier oil droplets impinge on the quarl and form carbon deposits.

Another important factor for the formation of the primary flame is that it must be supplied with primary air in the correct proportion and at the right velocity. In the case of air registers using high velocity air streams this is done by fitting a tip plate which spills the primary air over into a series of vortices, as indicated in Fig. 70. This ensures good mixing of the air and fuel and, by reducing the forward speeds involved, helps to maintain the primary flame within the refractory quarl.

SECONDARY FLAME

The larger oil droplets, heated in their passage through the primary flame zone, then vaporize and begin to burn. This, although a rapid process, is not instantaneous, and so it is essential that oxygen is supplied steadily and arranged to mix thoroughly with the burning particles of oil. An essential feature for the stability of this suspended secondary flame is that the forward velocity of the air and oil particles must not exceed the speed of flame propagation. If it does the flame front moves further out into the furnace and the primary flame will now burn outside the quarl with resulting instability due to overcooling.

92

As indicated in Fig. 70 careful design of the swirl vanes in the air register can be used to create the required flow patterns in the secondary air stream. The secondary flame gives heat to the surrounding furnace for the generation of steam.

Sufficient time must be given for complete combustion to take place before unburnt oil particles can impinge onto tubes or refractory material. This usually entails the supply of a certain amount of air in excess of the theoretical amount required for complete combustion if these practical considerations could be neglected, and unlimited time taken for the mixing of the air and fuel. The actual amount of excess air supplied depends upon a number of factors, such as the design of the furnace, the efficiency of the combustion process for the condition of load, etc., but will in general reduce the boiler efficiency to some extent due to the heat carried away by this excess air leaving the funnel. It can also lead to increased deposits in the uptakes due to the increased amount of sulphur trioxide that will form from sulphur dioxide in the presence of excess oxygen.

Q. Sketch and describe a pressure jet fuel oil burner. State how the throughput of oil is controlled.

A. A pressure jet oil burner, such as that shown in Fig. 71, forms a simple robust unit, widely used in marine boilers.

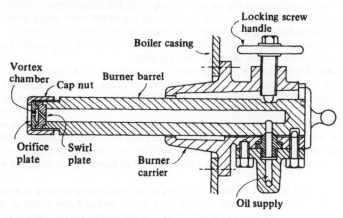

Fig. 71 Pressure jet type fuel oil burner

The basic assembly consists of a steel tube, or barrel, to which are attached swirl and orifice plates; these are made of a high grade or low alloy steel, and are held in place by a cap nut.

The complete unit is clamped into a burner carrier attached to the boiler casing. This both holds the burner in its correct position relative to the furnace, and also permits the supply of fuel through an oil tight connection. Some form of safety device must be fitted in order to prevent the oil being turned on when the burner is not in place.

The oil is supplied to the burner under pressure and, as it passes through, the burner performs two basic operations. First it imparts rotational energy to the oil as it passes through angled holes in the swirl plate. The rotating stream of oil thus formed

93

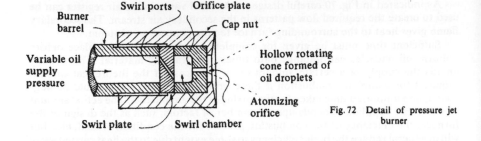

Fig. 72 Detail of pressure jet burner

is then forced under pressure through a small hole in the orifice plate which causes the jet to break up into fine droplets. This latter process is referred to as atomization, although each individual droplet of oil is formed of vast numbers of atoms.

As the final result of these operations a hollow rotating cone, formed of fine particles of oil, leaves the burner tip.

Many variations of design exist for the swirl and orifice plates. The choice of the actual design used often depends upon the means employed by the manufacturer to carry out the accurate machining processes required for these items.

In this type of burner control over the throughput of oil is obtained in two ways; by varying the oil supply pressure and/or by changing the diameter of the hole in the orifice plate. Limitations exist which prevent either method being used as the sole means of control over a wide range of throughput.

The ratio of the maximum to minimum oil throughput of the burner is known as the turn down ratio of the burner, and in the case of pressure jet burners this can be stated in terms of the square root of the ratio of the maximum to minimum oil supply pressures.

In all pressure jet burners, however, a minimum supply pressure in the order of 700 kN/m² is necessary to ensure efficient atomization is maintained. At the same time various practical considerations limit the maximum pressure to about 7000 kN/m², thus the turn down ratio with this type of burner is limited to a value of about 3·5.

If a wider range of turn down is required a system incorporating a number of burners is used, which controls the overall turn down on the basis of the number of burners in operation, or changing the orifice size in addition to the variation in supply pressure considered above. However, while this system is convenient for manual operation, it is not suitable for automatic control due to the need to change orifice sizes when the oil supply pressure reaches its upper or lower limits.

The burners must be kept clean and care should be taken during this operation not to damage or scratch the finely machined surfaces of the swirl and orifice plates. The latter should be renewed as the orifice wears beyond a certain amount. This should be checked at regular intervals by means of a gauge. After cleaning make sure all the various parts are correctly assembled. Any oil leaks must be rectified as soon as possible as they can lead to fires in the air register or double casing of the boiler.

Burners not in use should be removed, otherwise the heat from the furnace will cause any oil remaining in the burner barrel to carbonize.

Q. Sketch and describe a rotating cup type of fuel oil burner.

A. A rotating cup oil burner atomizes the oil by throwing it off the edge of a tapered cup being rotated at high speeds of between 2000–7000 rpm by either an air turbine driven by primary combustion air, or by an electric motor driving the cup shaft by means of vee belts.

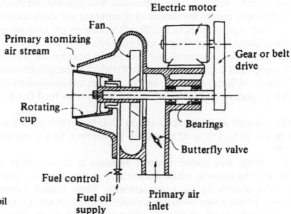

Fig. 73 Rotating cup type fuel oil burner

The basic assembly indicated in Fig. 73, consists of a tapered cup fitted onto the end of a central rotating spindle mounted on ball or roller bearings. The fuel oil is supplied to the inner surface of the cup through the hollow end of the spindle. Here centrifugal force causes it to spread out evenly into a thin film, which then moves out along the taper until it reaches the lip of the cup, where the radial components of velocity cause it to break up into fine particles as it passes into the surrounding air stream. Thus like a pressure jet burner this type of burner performs two functions: first, supplying rotational energy to the oil, and then breaking it up into fine particles. The final result is a hollow rotating cone of oil droplets leaving the burner.

High oil supply pressure is unnecessary as this pressure plays no direct part in the atomization process, and only sufficient pressure to overcome frictional resistance to the flow of oil through the pipes is required. Thus this type of burner can be used with a gravity type oil fuel supply system.

The oil throughput is controlled by a regulating valve placed in the fuel supply line, and thus can easily be adapted to automatic control. Here the wide turn down ratio available with this type of burner is a great advantage. Values of over 10: 1 are possible.

The diameter of the cup must be large enough to handle the required throughput, and there must be sufficient taper and rotational speed to ensure the oil is thrown off with the desired velocity. These factors govern the maximum oil throughput of the burner; the minimum throughput is limited only by the fact that sufficient oil must be supplied to maintain a continuous film of oil inside the cup so as to provide a stable primary flame.

The flame produced by a rotating cup burner tends to be long and cigar shaped, although a shorter flame can be obtained by careful design of the swirl vanes in the air register, so as to direct the flow of air in such a manner as to give the desired flame shape.

95

In the smaller units it is possible to supply all the combustion air through the burner itself, the air flowing through the space between the rotating cup and the fixed casing. However, in most cases only the primary air, which in this type of burner is used mainly for atomization, is supplied in this way. It only forms about 10 per cent of the total air required, the remainder being delivered through a secondary air register, to which it passes by means of a separate air duct with its own forced draught fan.

This type of burner is difficult to design for very large throughputs, and still give the required flame shape, and so while very suitable for auxiliary boilers with their relatively small outputs, they are not in general use for main water tube boilers. Here if more than one burner is to be fitted, the complication inherent in each rotating cup burner, with its own drive motor, makes other systems of atomization more suitable. Also rotating cup burners cannot be used in roof fired boilers.

Q. Sketch and describe a steam blast jet type fuel oil burner. Discuss the reasons for the use of this type of burner in preference to a pressure jet type fuel oil burner.

A. With automated control systems it is advisable to avoid extinguishing and re-igniting burners while manoeuvring, etc. It is also impracticable to change the size of the atomizing tip automatically. Thus simple pressure jet burners with their limited turn down ratios on a single orifice size are not suitable, since it is necessary to use a type of fuel oil burner with a large turn down ratio. Various forms of these wide range burners are available, and one type in common use is the blast jet burner.

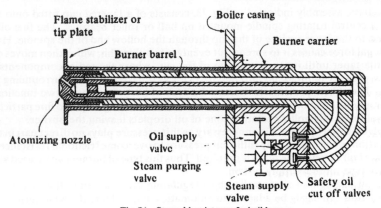

Fig. 74   Steam blast jet type fuel oil burner

These atomize the oil by spraying it into the path of a high velocity jet of steam or air. Although either medium can be used, steam is usually both more readily available and economical at sea. Compressed air is therefore seldom used, except when lighting up from cold.

Fig. 74 shows in diagrammatic form the general arrangement of a steam blast jet burner of the widely used Y-jet type.

In this the steam flows along the central passage, and is then expanded through a convergent divergent nozzle, where its pressure energy is converted to kinetic energy resulting in a high velocity jet of steam. Oil sprayed into this jet is entrained by it, being torn to shreds and atomized in the process.

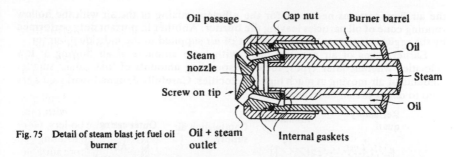

Fig. 75   Detail of steam blast jet fuel oil
          burner

The exit ports are arranged tangentially, thus giving the necessary swirl to the oil droplets in order to form the hollow rotating cone of fine particles of oil needed for the efficient combustion of a residual fuel oil in the boiler furnace. However, the flame shape is not so clearly defined as those obtained with pressure jet type burners due to the entrainment of air by the high velocity steam. This enables simple air registers to be used. There is no need to fit the usual swirl vanes for the secondary air stream—only a venturi shaped throat and tip plate are required.

The throughput of oil is controlled by varying the oil supply pressure. Since the atomizing effect is not obtained directly by the use of pressure, the same limit is not imposed on the use of very low oil supply pressures as with simple pressure jet burners; large turn down ratios of up to 20: 1 are therefore available with blast jet burners without having to resort to unduly high pressures. The oil supply pressure ranges from about 140–2000 kN/m², with corresponding steam pressures of 140–1500 kN/m².

Care must be taken to use only dry steam, any water present having a chilling effect which could cause flame instability. The steam may be obtained directly from the boiler, the pressure being dropped to the required value by passing it through reducing valves. Alternatively it may be obtained from an auxiliary source such as a steam to steam generator.

Excessive use of steam can be caused by incorrect setting of the burner, or by leakage across the joint faces in the atomizing head of the burner, and in some versions gaskets are fitted to prevent this as shown in Fig. 75.

Steam is left on all the time the burner is in operation, even when the oil is turned off, in order to cool the burner and prevent any remnants of oil in the burner passages from carbonizing.

Safety shut off valves are fitted to the burner carrier; these are opened by projections on the burner so that oil and steam are automatically shut off when the burner is removed.

Q.   Sketch and describe an air register suitable for the supply of combustion air to the furnace of a water tube boiler.

A.   The term air register is applied to the assembly of vanes, air swirler plates, etc. fitted within the double casing of the boiler in way of each burner position, in order to supply the air required for combustion in the correct manner. The width of the casing at this point is determined by the design of the register used.

The register performs the following functions: it divides the incoming combustion air into primary and secondary streams, and then directs these streams so as to give the air flow patterns necessary for the efficient mixing of the air with the hollow rotating cone of oil particles leaving the burner. Another important duty performed by the register is to regulate the amount of air supplied to the individual burner.

Earlier types of air registers dealt with large amounts of air flowing at low velocities whereas later types, for corresponding amounts of fuel, admit smaller amounts of air moving at much higher velocities. Carefully designed swirl vanes are used to direct the air as required.

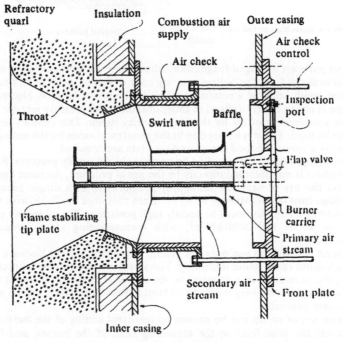

Fig. 76   Air register

Many variations in design exist, and here any constructional details apply only to the register shown in Fig. 76, but some, or all of the components shown, are common to all types of air register.

The combustion air must pass through the air check in order to enter the register. In some cases the check is formed by the swirl vanes themselves being rotated about their axes until they touch, so shutting off the air flow to the burner. However, in most cases some form of sliding sleeve is used. This type of check is shown in Fig. 76.

The air check may be operated by hand, usually being placed in a fully open or fully closed position. However, in the case of wide range burners, especially where automated combustion control systems are fitted, a pneumatic means of operation is used. In some cases the check may be placed in an intermediate position in order to adjust the air supply to individual burners.

Baffles are then often fitted to separate the air into primary and secondary air flows; these are concentric within the cylindrical register. The inner primary air stream flows along until it reaches the tip plate fitted at the end of the burner tube.

Here the air impinges on the back of the plate, and then spills over to form a series of vortices which have the effect of reducing the forward velocity of the air, and so helps to retain the primary flame within the quarl. Various designs of tip plates are used, ranging from circular flat plates to more complex swirl vanes, but all perform the same basic function; i.e. the formation of the vortices which is so important in modern air registers with their high velocity air flows.

The outer, secondary air stream passes over swirl vanes which cause it to rotate as it passes through the quarl, so giving better conditions for the mixing of fuel and air in the secondary flame zone. Again by careful design the air flow pattern can be made to form a series of vortices. In this way the forward velocity of the burning oil particles is reduced, giving a longer period for combustion to take place within the furnace. The secondary air flow also has some influence on the flame shape; this is especially the case with rotating cup type burners.

A small amount of cooling air is often allowed to flow between the inside edge of the tip plate and the atomizing tip of the burner. This amount of air must remain small, otherwise it can upset the vortex system formed by the tip plate.

It is important that the air check forms a tight seal in order to prevent combustion air entering the register when the burner is not in use, otherwise thermal shock caused by the relatively cool combustion air leaking through can damage the refractory quarles.

An insulated front plate must be fitted in some cases to prevent overheating of the boiler front due to radiant heat from the furnace penetrating through the gaps between the swirl plates, etc.

In some types of registers, especially those used with simple pressure jet burners with their small turn down ratios, very little adjustment of the relative positions between the various vanes, air swirler plates, etc. can be carried out. In others a whole range of minor adjustments may be carried out to suit different fuels, conditions of load, etc.

It may be noted that, in the case of steam jet burners, the steam provides additional energy for the mixing of the air and fuel, and the swirl vanes for the secondary air stream may be omitted from air registers used with this type of burner.

Q. Describe a fuel oil system from the settling tanks to, but not including, the burners.

A. Fig. 77 shows the diagrammatic layout of a semi-automatic fuel oil system which makes use of wide range burners operating on a variable oil supply pressure. The basic layout consists of a ring main supplying the individidual burners by a series of dead legs.

Oil is pumped into the settling tanks, and any water allowed to settle out. This can then be drained off by means of spring-loaded drain cocks.

When required for use, the high suction valve on the settling tank is opened and oil allowed to pass to the cold filters fitted on the suction side side of the fuel oil service pump.

Due to the high viscosity of the unheated residual oil only a coarse filter, just sufficient to prevent damage to the pump, can be used at this stage. This consists of a positive displacement pump operating at a constant delivery pressure. A spring-loaded relief valve fitted on the discharge side of the pump allows any excess oil to spill back to the suction side of the pump in the event of overpressure.

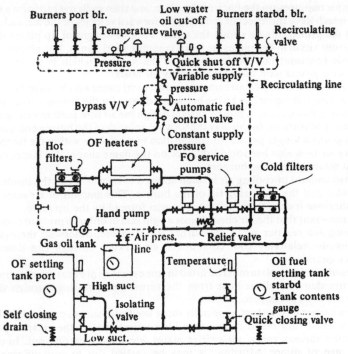

Fig. 77 Fuel oil system

After leaving the pump, the oil temperature is raised in a fuel oil heater. This is done in order to lower the viscosity of the oil, making it easier to filter and finally to atomize. The correct oil temperature is maintained by means of a thermostat placed in the outlet from the fuel oil heater, which controls the supply of steam to the heater.

The hot filters fitted after the heater are normally of an auto-clean type; in some cases they are constantly rotated by electric motors. These filters provide a fine filtration. This prevents wear and chokage of the fine passages in the atomizing tip of the fuel oil burners.

The heated and filtered oil now passes through an automatic pneumatically-operated control valve which varies the oil supply pressure to the burners in response to variations in the main steam pressure transmitted to a master pressure controller. The combustion air controller also varies in order to maintain the correct ratio between the amounts of fuel and air supplied to the furnace.

Two emergency valves are now fitted; the first is a manually-operated quick shut off valve which enables the fuel oil to be shut off by hand from the boiler very rapidly in case of emergency. The second is a shut off valve with a steam actuator which operates to shut off the fuel oil in the event of loss of water in the boiler.

The oil is now ready to enter the individual dead legs supplying oil to the burners. An isolating valve and a safety cock, or similar device, is fitted to each leg.

To enable the oil temperature in the system to be brought quickly up to and then maintained at the desired operating value, a recirculating valve is fitted which enables oil to be circulated through the ring main back to the pump suction. This valve is closed as the burners are brought into operation.

In the system shown in Fig. 77 the individual burners are ignited by means of a paraffin torch, but in many cases automatic igniting devices are fitted. In this case it is advisable that a flame failure alarm should also be fitted.

The system uses gas oil to flash up from cold. Air pressure supplied to the gas oil storage tank is used to force the oil through to the burner. The use of this oil continues until sufficient steam has been generated to enable the fuel oil heaters to be put into operation so as to raise the temperature of the residual fuel oil to the value required for its combustion.

All necessary pressure gauges, thermometers, air vents, etc. must be fitted for the proper operation of the system.

Safety fittings include quick closing valves—operated from outside the engine-room—fitted to the suction lines from the settling tanks, which enable fuel to be shut off from the system in case of emergency. There is also an emergency stop fitted to the fuel oil service pump.

All oil lines and fittings containing heated oil should be placed above the plates, in well-lit situations, so that any leakage can be easily detected.

The system shown can easily be changed to manual control by simply bypassing the automatic control valve, and controlling the oil supply pressure by means of a hand jacking wheel on the spring-loaded relief valve governing the discharge pressure from the fuel oil service pump.

Q. Discuss the use of refractory materials in boilers.

A. The basic requirement of boiler refractory materials is that they should contain the heat generated in the furnace. They must therefore have good insulating properties and be able to withstand the high temperatures to which they will be exposed. They must also have sufficient mechanical strength to resist the forces set up by the weight of adjacent brickwork, etc.; to withstand vibration; and the cutting and abrasive action of the flame and flue dust. The materials must also be able to expand and contract uniformly without cracking.

At the present time no single refractory material can be used economically throughout the boiler, and the choice of suitable materials for various parts of the boiler is generally governed by the temperatures to which they will be subjected.

The material from which these refractories are manufactured can be grouped into three main types:

Acid materials: Clay, silica, quartz, sandstone, gamister.
Neutral materials: Chromite, graphite, plumbago, alumina.
Alkaline or base materials: Lime, magnesia, zirconia.

It should be noted when choosing suitable materials that care must be taken to ensure acid and alkaline substances are kept apart as, at high temperatures, they can react to form salts which destroy the effectiveness of the refractory.

These refractory materials are available for installation in one of two basic forms:

FIREBRICKS

These are formed into bricks and then fired at high temperatures in special kilns.

MONOLITHIC REFRACTORIES

These are supplied in an unfired state, installed in the boiler, and fired in situ when the boiler is put into service. This form of refractory can be subdivided into:

Mouldable refractory This is used where direct exposure to radiant heat takes place. It must be pounded into place during installation.

Castable refractory Placed behind water walls and other parts of the boiler where it is protected from direct exposure to radiant heat. It is installed in a similar manner to building concrete.

Plastic chrome ore This, bonded with clay, is used in the construction of studded water walls. It can resist high temperatures, but has little mechanical strength, and is pounded onto steel studs welded to the tubes. These studs provide both strength and means of attachment for the refractory.

All the forms of refractory materials previously mentioned must be securely attached to the boiler casing and, in addition to the above-mentioned studded tubes, various types of bolts, clips, and keys are used for this purpose.

To prevent undue stresses being set up in the refractory, ample expansion spaces must be provided. Care must be taken to ensure these spaces do not become blocked in any way as this can cause the refractory to break away from its attachments and bulge out, with the danger of possible collapse.

Refractory materials impose limits upon the time required for raising steam; the greater the amount of refractory, the slower the steam raising process must be in order to prevent damage to the refractory material.

Air checks should be closed immediately the corresponding burner is shut off, otherwise the relatively cold air impinging upon the hot refractory causes a surface flaking, known as spalling, to take place. This leads to a reduction in wall thickness.

Flame impingement must be avoided as it leads both to a build-up of carbon deposits, and to damage caused by carbon penetrating into the refractory.

Another form of damage is caused by impurities in the fuel: mainly vanadium and sodium salts reacting with the refractory material to form a molten slag, which then runs down to the furnace floor. This both reduces the wall thickness, and leads to a build-up in the level of the furnace floor which can eventually interfere with the shape of the flame.

Q. Sketch and describe a soot blower suitable for fitting to a water tube boiler. State why they are fitted.

A. In order to maintain the gas side heating surfaces of a boiler in a clean condition and so prevent an undue build-up of deposits which can lead to corrosion and/or uptake fires, soot blowers are fitted.

Soot blowers consist of two main parts, the head or chest, and one or more nozzles attached to a tube or spindle made of heat-resistant steel. An operating mechanism fitted in the chest rotating and, in some cases, retracting the nozzles as required.

The efficiency of the blower depends upon the conversion of pressure energy in the blowing medium to kinetic energy; this results in a high velocity jet of fluid impinging upon the sooted surfaces.

Although in a few cases the soot blowers operate with compressed air, the majority of marine soot blowers use steam.

Superheated steam is normally used to maintain dry conditions, and also to ensure that the superheater is not starved of steam while blowing is in progress.

Soot blower steam lines should be sloped so as to be self-draining to a suitable drain valve. These are often automatic to ensure they remain full open when the steam is shut off, closing when the steam is turned on, but not until sufficient steam has been allowed to blow through in order to warm the lines.

The reason for leaving the drain full open, when the master valve supplying steam to the soot blowers is closed, is to prevent any leakage past this valve leading to a build-up of pressure, which would force steam to leak past the blower seal rings and so cause corrosion of the blower nozzles.

When air or high pressure steam is being used, a double shut off valve is fitted in addition to the seal rings to ensure the air or steam only enters each soot blower as it actually commences its blowing sequence.

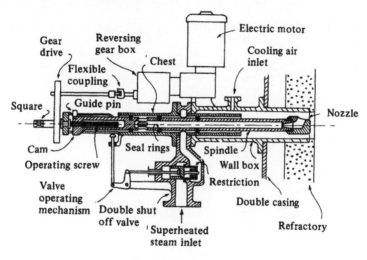

Fig. 78   Soot blower

The general arrangement of a high pressure, automatically operated soot blower is shown in diagrammatic form in Fig. 78. It is operated by an electric motor driving through a reversing gear box. The motor is started in its correct sequence from a central control consol, moving the blower through its correct blowing procedure until a cam arrangement in the gear box switches it off, and as it does so it gives a signal for the next blower to start.

A handle can be fitted to the square at the end of the operating screw to enable the blower to be operated manually.

The type shown is retractable, so first the operating screw extends the nozzle to its blowing position. As it comes towards the end of its travel, the cam opens the double shut off valve, allowing steam to pass through to the blower chest, where the seal rings have now moved to a position which allows the steam to enter the admission ports in the sliding spindle. The steam passes along this spindle to the convergent

103

divergent nozzle where the expansion process taking place results in a high velocity steam jet.

A guide pin moving in a slot causes the nozzle to be rotated through its blowing arc while the steam is on.

To prevent high pressure steam eroding tubes in high temperature regions of the boiler, restriction or orifice plates, fitted in the steam supply passage to the individual soot blowers, reduce the pressure as required.

Cooling air is sometimes supplied to the wall box tube. This air is normally taken from the forced draught fan discharge before the air heater.

CHAPTER 7

Boiler Operation

Q. Describe the process of raising steam from cold on a Scotch boiler.

A. If the boiler has been opened up for cleaning or repairs check that all work has been completed, and carried out in a satisfactory manner. Ensure that all tools, etc. have been removed. Examine all internal pipes and fittings to see that they are in place, and properly fitted. Check that the blow down valve is clear. Then carry out the following procedure:

Fit lower manhole door. Check external boiler fittings to see they are in order. See that all blanks are removed from safety valves, blow down line, etc.

Fill boiler with water to about one-quarter of the water level gauge glass. If possible hot water heated by means of a feed heater should be used. The initial dose of feed treatment chemicals, mixed with water, can be poured in at the top manhole door at this stage if required. Then fit top manhole door. Make sure air vent is open.

Set one fire away at lowest possible rate. Use the smallest burner tip available. By-pass air heater if fitted. Change furnaces over every twenty minutes.

After about one hour start to circulate the boiler by means of auxiliary feed pump and blow down valve connection, or by patent circular if fitted. If no means of circulation is provided, continue firing at lowest rate until the boiler is well warmed through especially below the furnaces. Running or blowing out a small amount of water at this stage will assist in promoting natural circulation if no other means is available. Continue circulating for about four hours, raising the temperature of the boiler at a rate of about 6°–7°C per hour. Water drawn off at the salinometer cock can be used to check water temperature below 100°C.

At the end of this time set fires away in all furnaces, still at the lowest rate. Close the air vent.

Nuts on manhole doors, and any new joints should be nipped up. Circulating the boiler can now be stopped, and steam pressure slowly raised during the next 7–8 hours to within about 100 kN/m² of the working pressure.

Test the water gauge. The boiler is now ready to be put into service. About 12 hours should be allowed for the complete operation provided some means of circulating the boiler is provided. If circulation cannot be carried out, the steam raising procedure must be carried out more slowly, taking about 18–24 hours for the complete operation. This is due to the fact that water is a very poor conductor of heat, and heat from the furnace will be carried up by convection currents leaving the water below the furnace cold. This will lead to severe stresses being set up in the lower sections of the circumferential joints of the boiler shell if steam raising is carried out too rapidly, and can lead to leakage and 'grooving' of the end plate flanging.

If steam is being raised simultaneously on more than one boiler, use the feed pump to circulate each boiler in turn, for about ten minutes each.

Q. Describe the procedure for opening up a Scotch boiler. What inspections should be carried out before the boiler is again boxed up?

A. Empty the boiler, preferably by allowing the boiler to cool down, and then running or pumping out. If there is not sufficient time for this, allow boiler pressure to fall to 300–400 kN/m² and blow down. When pressure is off the boiler, open the air vent and allow the boiler to cool down.

When the boiler is cool, make sure there is no vacuum in the boiler; this should be done by opening the drain cock on the water level gauge glass in case the air vent is choked.

Then commence to open up the boiler by first removing the top manhole door. To do this, slacken back the nuts holding the dogs, but do not remove them until first breaking the joint. This precaution should be taken in the event of pressure or vacuum existing in the boiler. The nuts and dogs can then be removed, and the door removed. Depending upon the weight of the door, it may be necessary to rig a lifting block to the door in order to do this.

The opening should then be roped off, and all personnel warned to keep clear. The bottom door can now be removed, again taking care when breaking the joint in case water is still above the sill of the door. If this should be the case, pump out before removing door.

It is important that this sequence be followed as, when the lower door is removed, it allows a through-draught and hot vapour rising through the top door may scald anyone standing over the hole. Hot vapour can remain in a Scotch boiler even after a considerable period of time allowed for cooling down.

With the doors removed, allow the boiler to ventilate before attempting to enter. Do not allow naked lights near the boiler until it has ventilated due to the danger of explosive gas in the boiler. If in doubt, use a safety lamp to test the atmosphere in the boiler is safe to breathe before entering.

A preliminary internal inspection should be carried out before cleaning is commenced to check the general condition. Note scale deposits and any special points.

Plug the orifice to the blow down valve to ensure it does not get choked during cleaning operations, and place guards over the manhole landings to ensure they are not damaged. The boiler can now be cleaned, and any internal work carried out.

When all work is completed, a full internal examination must be carried out. It is advisable to keep a record of the boiler, consisting of a drawing on which any troubles, repairs, etc. can be shown, and a book in which remarks regarding scale formation, corrosion, deformation, etc, can be kept.

Check to see all cleaning has been carried out efficiently, especially where the tubes enter the tube plates. See that all tools and other articles have been removed from the boiler, paying special attention to combustion chamber top, tube nests, and bottom of boiler. Make sure all openings are clear, taking special care with the water level gauge connections to ensure they are clear and free from deposits. Make sure all internal pipes and fittings have been replaced correctly, and are securely attached. The guards can be removed, and the faces of the manhole doors and landings inspected to see they are clean and undamaged.

Remove the plug from the blow down valve orifice. Replace the lower manhole doors, using a new joint. Operate all boiler mountings and see they work correctly. Leave in a closed position, except for water level gauge steam and water cocks, and air vents.

The boiler can now be filled to one-quarter level in the gauge glass if steam is to be raised, or filled completely if a hydraulic test is to be carried out.

Q. State the regulations concerning the hydraulic testing of a Scotch boiler. Describe how you would carry out such a test.

A. New boilers having a design pressure in excess of 690 kN/m^2, together with their components, must be subjected to a hydraulic test at a pressure = (1·5 × design pressure + 350) kN/m^2 upon completion. For boilers working at pressures below this value the test value is 2 × design pressure.

The test must be carried out in the presence of an authorized surveyor, who upon satisfactory completion of the test will stamp the boiler with the official DOT stamp if it is for a passenger vessel, or if a classification society surveyor is concerned, their official stamp will be used. The surveyor's initials are also put on alongside the stamp, which is usually on the bottom front plate, near the furnace.

Boilers which have undergone structural repairs must be subjected to a hydraulic test at a pressure at least equal to the design pressure.

The surveyor may call for a hydraulic test at any survey, the test pressure being to the surveyor's requirements.

The procedure for such a test is carried out as follows. Close or blank off all openings. Measuring tapes may be placed around the boiler, and deflection gauges in the furnace. Lagging should be removed as required to facilitate inspection of joints, etc.

The boiler is then completely filled with water, the air vent being left open until water shows to ensure no air is trapped inside. It should be noted that the use of hot water places the boiler closer to working conditions, but may scald in the event of failure if the water is hot enough to flash off into steam with the resultant drop in pressure.

The force pump, and test gauges can now be fitted. The gauge glasses should be shut off if the test pressure is to be above the design pressure. The readings on the measuring tapes, and deflection gauges should be noted. The boiler can now be pressurized by means of the force pump. Care should be taken to ensure that the pressure rises smartly in response to the pumping action; if it appears sluggish, open the air vent to remove any air remaining in the boiler. Listen carefully during application of pressure in case any combustion stays, etc. fracture.

Then examine all joints, especially if these are riveted. Check flanges for cracks. All flat surfaces should be checked with a straight edge for signs of bulging due to stay failure, overheating, or thinning of the plate. Look for signs of leakage at tell-tale holes in the combustion chamber stays and welded compensating rings. Examine all tube ends for signs of leakage. Check, and note readings on measuring tapes, and deflection gauges.

The test pressure must be maintained until the surveyor has completed his examination, and must in any case be kept on for at least ten consecutive minutes. The pressure can then be released. Readings on the measuring tapes and deflection gauges should again be checked to ensure they have returned to their initial values. The boiler can then be emptied, and examined inside and out.

Q. Describe a procedure for closing up, and then raising steam on a water tube boiler.

107

A. Before closing up the boiler inspect the internal surfaces to ensure they are clean, all openings to the boiler mountings clear, and tubes proved to be free of obstruction by means of search balls, flexible wires, air or water jets.

Replace any internal fittings which have been removed, checking to ensure they are correctly positioned and secured.

The header handhole plugs and lower manhole doors are now replaced.

Operate all boiler mountings to ensure they work freely, leaving all the valves in a closed position.

Check the gas side of the boiler is clean and in good order. Make sure the soot blowers are correctly fitted, and operate over their correct traverse. Operate any gas or air control dampers fitted to ensure they move freely for their full travel. Leave them closed or in mid-position as necessary. The boiler casing doors are now replaced.

Open the direct reading water level gauge isolating cocks, together with all boiler vents, alarm and pressure gauge connections. The superheater drains are also opened. Check that all other drains and blow down valves are closed.

Commence to fill the boiler with hot deaerated water. At this stage the initial dose of chemical treatment can be added through the top manhole doors, which are then replaced.

Continue to fill until water just shows in the water level gauges. Close any header vents as water issues.

Remove the funnel cover, and ensure that all air checks operate correctly and that the forced draught fans are in working order. If gas air heaters are fitted they should be by-passed.

Check the fuel oil system to ascertain it is in good order. Start up the fuel oil service pumps and check for leaks. The boiler is now ready to commence raising steam.

Heat the fuel oil to the required temperature, using the recirculating line to get the heated oil through the system. If no heat is available for this, use gas oil until sufficient steam is available to heat the residual fuel oil normally used.

Start the forced draught fan, and with all the air checks full open purge the boiler, making sure any gas control dampers are in mid-position so giving a clear air passage.

Carry out a final check to make sure water level gauge cocks are open, water is showing in the glass, and that steam drum and superheater vents are open.

Now close all the air checks except for the burner to be flashed up, this being done by means of ignition equipment or a paraffin torch. Use the lowest possible firing rate. Adjust the air supply so as to obtain the best combustion conditions and check that, as the boiler heats up, the water level in the glass begins to rise.

After about one hour steam should show at the drum and superheater vents and, when issuing strongly, open the superheater circulating valve and close the air vents.

When the steam pressure has reached a value of about 300 kN/m^2 blow through the water level gauges to ensure they are working correctly. The isolating valves on the remote reading water level indicator can now be opened, and the indicator placed in service.

With the steam pressure at about 1000 kN/m^2 follow up the nuts on all new boiler joints.

At a pressure of about 1400 kN/m^2 open the drains on the auxiliary steam lines, crack open the auxiliary stop valve and warm the auxiliary line through. Now close the drains and fully open the auxiliary stop valve.

Various auxiliary equipment such as fuel oil heaters, turbo-feed pumps, etc. can be put into service and, provided this entails a flow of steam through the superheater, the superheater circulating and drain valves are closed.

Bring the boiler up to working pressure, keeping the firing rate as steady as possible, and avoiding intermittent flashing up.

Check the water level alarms.

Open the main steam line drains, and crack open the main stop valve and warm through the main steam line. Then close the drains and fully open the main stop valve.

The procedure from flashing up to coupling up at full working pressure should take about four to six hours. Only in emergency should it be carried out more rapidly. If new refractory material has been installed carry out the procedure more slowly.

At all times during the raising of steam the superheaters must be circulated with steam to prevent them overheating. If the temperature of the superheaters goes above the permitted value for the boiler reduce the rate of firing.

It must be noted that, due to the great variety of water tube boiler designs in use, the foregoing procedure is only to be taken as a guide; for example, header boilers with their greater amount of refractory material will require about eight hours to reach full pressure. Thus the engineer should always follow the procedure laid down for his particular boiler, which may vary in detail from the basic principles previously stated.

Q. Describe the procedure for carrying out a hydraulic test on a water tube boiler. To what pressure would you subject the boiler for purpose of testing.

A. New boilers having a design pressure in excess of 690 kN/m^2 together with their components must be subjected to a hydraulic test at 1·5 times the approved design pressure, carried out to the surveyor's requirements.

Although the surveyor may call for a hydraulic test at survey, it is not normally required for routine survey unless limited access prevents a practicable visual examination. After repairs to pressure parts a hydraulic test to 1·25 times working pressure is usually accepted; for minor repairs such as tube renewal, testing to just below working pressure will suffice.

For purpose of testing the boiler is considered to extend from economiser inlet to main steam stop valve.

Before testing, carry out a thorough inspection of the boiler to ensure that all necessary work has been completed, all internal surfaces are clean, and all tools etc. have been removed. Remove any casing as necessary for a proper inspection of the pressure parts undergoing test. In some instances it may be necessary to remove some of the boiler insulation.

Gag the safety valves and shut off the water level indicators, alarms etc. if going above the working pressure, and isolate or blank off any parts not designed to withstand the test pressure. Pay particular attention to attemperators or de-superheaters, as these are often designed to cope only with the differential pressures between the drum and superheater; it may be necessary to equalise pressures across them during the test, in some cases by slacking back one of the flanged connections.

Check that all valves operate freely and seat properly. Then close all valves, drains etc. on the boiler but leave all vents open. All manhole doors can be closed, and any header plugs replaced.

Fit the force pump and any test gauges required, ensuring that these are properly calibrated.

Start to fill the boiler using water as close to the metal temperature as possible, but not less than 7°C, as any sudden change in temperature such as may be caused by filling a warm boiler or superheater with cold water may cause leakage by way of expanded joints. The boiler must also be protected against mechanical or thermal shock during the test, so never put a hydraulic test on a hot boiler or superheater. During filling, check for leaks, open drains etc. and if any are found stop filling until they are rectified. Close vents as water issues from them, continuing to fill until the boiler is completely pressed up, and all air is released.

The force pump can now be used, building up the pressure slowly so as to avoid shock. If the pressure does not begin to rise with the first few strokes of the pump, check all vents to ensure that all air has been released.

When fully pressurised, check all visible seams, expanded joints etc. for signs of leakage or distortion, maintaining the test pressure for at least 30 minutes.

With inspection completed, slowly release the test pressure. If any leaks have been detected, the water level need only be dropped sufficiently for the leak to be recified. Retest until no leakage occurs.

Before draining the boiler it is a convenient time, after the gags have been removed, to use hydraulic pressure to reset the safety valves if required. They will then only need a final adjustment under steam.

Should raw water have been used for testing, completely drain and then flush out the boiler with distilled water before returning to service.

Q.　Give the basic survey procedure to be carried out on a marine boiler.

A.　Surveys are carried out to ensure that the boiler is in a safe working condition and likely to remain so until the next survey. Main water tube boilers of passenger ships are surveyed annually, while for cargo ships bi-annual survey is sufficient. The normal procedure for auxiliary boilers is bi-annual for the first eight years, thereafter annual, although a concession may be made for auxiliary water tube to continue on a two year cycle, provided that they are in good condition and with good records, especially those for water treatment.

The survey will cover the boiler from burner front, or exhaust gas inlet, to funnel top, including all pressure containment parts, valves and fittings.

During the survey the boiler, economiser and air heater are to be examined both internally and externally as far as access permits, and where considered necessary pressure parts subjected to a hydraulic test. Thickness of plates and tubes is to be determined, usually by ultrasonic test equipment.

The principal boiler mountings are to be examined externally, and opened up for internal inspection where considered necessary. Boiler casings, supports etc. will be examined to see they are in good order, and allow for free expansion. Automatic control equipment for water level and for fuel oil is to be checked, with special regard to safety cut-outs.

Isolate the boiler from all working systems, and open up both water and gas sides, removing any internals necessary for a proper visual examination. A visual inspection should be made to ascertain general boiler conditions, before it is cleaned both internally and externally.

Previous records, if any, should be examined and note taken of any previous defects or repairs, so these can be given special attention. A convenient survey procedure taking into account the layout of the boiler should be decided upon. This will vary from one boiler to another, but a typical procedure will be as follows. The steam drum is examined internally and externally together with its mountings, then all top headers and the burner positions, including carriers and associated pipework. Economiser and superheater headers can be inspected before entering the water drum. Under the boiler, any mountings and supports are checked.

Do not carry out any gas side inspections until all internal examinations are complete, in order to avoid carrying grease, dirt etc. into the water spaces.

On the gas side, start with the furnace (staging may be required in large roof-fired units). Then in turn the superheater, economiser, and air heater can be inspected.

A note pad with pencil securely attached, an unbreakable torch and a mirror on a rod should be available, together with other items if desired such as a straight-edge, an introscope to examine tube bores, and a polaroid camera to record any defects found during the examination. In the case of welded boilers with limited access, blanket radiography pictures using portable gamma ray equipment may be obtained, and used to look for suspected internal foreign objects inside tubes etc.

The survey is not complete until the boiler has been examined under steam, and the pressure gauges checked against a test gauge. The water level indicators and protective devices must be tested, and the safety valves adjusted to their correct blow-off pressures. The fuel oil system is to be checked under pressure, and remote operated fuel oil cut-off gear tested.

In the case of exhaust gas boilers where steam cannot be raised in port, the ship's Chief Engineer will be responsible for the correct setting of the safety valves at sea, with the boiler survey record not being completed until he confirms that this has been done satisfactorily.

Q. Discuss some of the common problems that can arise in the operation of marine boilers.

A. Boiler damage can be considered under five main headings: corrosion, erosion, overheating, cracking and mechanical damage.

Corrosion   There are two principal forms of corrosion. One is direct chemical attack and mainly occurs in superheaters due to the high metal temperatures involved. It can result in pitting or cracking in tube bores, or in scaling or flaking on the gas side of tubes. This form of corrosion also occurs when loss of water circulation causes the metal to overheat in the presence of steam.

The more comon form of corrosion found in boilers is the result of electro-chemical attack usually involving acidic water conditions in the presence of dissolved oxygen. General wastage of the boiler metal due to this form of attack has been virtually eliminated by the use of chemical feedwater treatment, but isolated pitting can still occur if the treatment is not operated within the correct limits. It may be found along the water level in the drum, generally as the result of poor shut-down and storage procedures where the boiler is left partly filled with cold water. Pitting along the roof of the drum can result from condensation. In the lower parts of the boiler it can be due to poor drainage, pools of water remaining in drums and headers. However, isolated pitting can also result from operating with the boiler water allowed to remain at too low a pH value, and in this case will also tend to occur in the bores of tubes subjected to the highest rates of heat exchange, such as screen and water wall tubes.

Erosion   This is a mechanical wearing away of the boiler metal due to water, steam or gas flowing over the metal surface. Thus tubes can wear thin in the region of bends, due to water impingement, or wear externally as they stand in the flow of hot abrasive gases leaving the furnace. Serious local erosion can result from the direct impingement onto the tube surface of a steam jet from a badly aligned soot blower.

Overheating   In-service boiler metal subjected to the heat of combustion must be continually cooled by water or steam. If for any reason this cooling affect is lost, or greatly reduced, the boiler metal overheats, loses strength and distorts. This can result in expanded tubes pulling out of tube plates, local bulging of tube surfaces with eventual rupture, the sagging of superheater tubes between their supports, etc.

Loss of water brings about the most immediate and serious damage, but loss of circulation through tubes will also quickly result in damage. A build-up of deposits on the water side acts as an insulating layer, reducing the rate of heat transfer through the metal so causing it to overheat and leading to eventual distortion. Oil entering the boiler only forms a thin, but efficient, insulating layer upon heated surfaces, but also encourages a further build-up of scale deposits.

Superheaters at their operational metal temperatures are very vulnerable to overheating, and a circulation of steam through them must be provided at all times when the boiler is steaming, circulating to atmosphere if not load is available. Priming and carry-over must be avoided, as impurities passing over with the water will result in a build-up of deposits in the superheater tubes, again causing eventual overheating.

Direct flame impingement onto water walls will lead to overheating and distortion. Fire in uptakes or superheaters due to a build up of deposits in the gas passages can also result in serious overheating and damage even to pressure parts of the boiler.

Cracking   Welded boilers are especially vulnerable to fatigue cracking resulting from bad design, poor workmanship or both. Cracks of this nature, even if starting in a minor weld, can continue to propagate even into the main shell plate.

Fatigue cracking is also associated with thermal cyclic stressing, which can result from poor steam-raising procedures, lack of expansion, or even from a continual carry-over of water droplets causing cracking inside superheater headers.

Over-expansion of tubes into tubeplates can lead to cracking of the bellmouthing and in some cases to cracks forming in the tubeplate between the holes.

Cracking due to caustic embrittlement can take place in boilers of riveted construction, due to slight leakage allowed to continue over a period of time in a riveted seam.

Mechanical Damage   This can result from poor workmanship, such as damage to tube plates by over-expanding during tube attachment, scoring of joint faces, distortion of doors by overtightening.

Another source of damage are explosions in the furnace due to bad flashing-up procedures. These can be especially serious in roof-fired radiant heat boilers with gas tight water wall panels; the force of the explosion acting directly upon these can cause them to suffer severe distortion. It is normal practice to warm this type of furnace with heated air etc. before attempting to flash up from cold.

Q.   State the general precautions to be followed by a watchkeeper in charge of a water tube boiler installation.

A. The watchkeeping officer must familiarise himself with the plant of which he is in charge, and,be aware of all individual equipment operating control signals, flow rates, temperatures and general load conditions. He must check these regularly so as to become aware quickly of any deviations from the norm. Rarely do emergency conditions arise without some previous indication, which an alert watchkeeper should recognise, investigate, and then take corrective action before the situation gets out of hand.

Ensure that all boiler and associated safety shut-down devices are maintained in full operational condition, and tested at regular intervals so as to be ready for instant operation. All alarm and automatic control systems must be kept within the manufacturer's recommended operating limits. Do not allow equipment to be taken out of operation for reasons which could reasonably be rectified.

All control room check lists must be kept up to date, with any known deviations from normal operating procedures noted, both for immediate reference and to inform oncoming watchkeepers of the ongoing situation. Remember that any deterioration in watchkeeping standards can give rise to circumstances whereby deviations remain un-noticed and may build up to potentially serious conditions.

Automatic control loops do not think for themselves, and subjected to external irregularities will still try to perform as normal. This can result in their final control action being incorrect, or to some other piece of equipment being overworked in an attempt to compensate. In situations where the automatic control of critical parameters is not dependable, or where it becomes necessary to use manual control, reduce operating conditions so as to increase acceptable margins of error.

High performance water tube boilers demand high quality feed water, so do not tolerate any deterioration of feed water conditions; immediately trace the source of any contamination, and rectify the fault.

Do not neglect leakage of high pressure, high temperature steam, as even minor leaks will rapidly deteriorate. No attempt should be made to approach the site of leakage directly, but the defective system should be shut down as soon as is practicable and the leakage rectified. Do not allow steam and water leaks to go un-corrected as, apart from reduction in plant efficiency, they also lead to increased demand for extra feed with an inevitable increase in boiler water impurities.

Always be alert for conditions which increase the potential fire risk within the engine room: the best method of fire fighting is not to allow one to start. Thus all spaces, tank tops etc. must be kept clean, dry, and well lit. This not only improves the work environment, but also makes for the early detection of any leakage and encourages early repair.

Store any necessary stocks of combustibles remote from sources of ignition. Maintain all oil systems tight and free from leaks and overspills.

Follow correct flashing-up procedures for the boiler at all times, especially in the case of roof-fired radiant heat boilers.

Be familiar with the ship's fire fighting systems and equipment, and ensure that all under your direct control are kept at a full state of readiness at all times.

Assess particular risk areas, especially in engine room spaces, and formulate your approach in case of emergency; decide in some detail how you would deal with fires at various sites in the engine room.

Make sure that your are familiar with the quick closing fuel shut-off valves, the remotely operated steam shut-off valves etc. to enable the boiler to be put in a safe condition if having to abandon the machinery spaces in the event of a fire.

Q. State a basic procedure to be followed for the cleaning of a boiler after a period of service.

A. The frequency of boiler cleaning depends upon various factors such as the nature of the service in which the vessel has been engaged, the quality of feed water and fuel with which the boiler has been supplied. In general, every reasonable opportunity should be taken, whenever the boiler is shut down, to examine internal and external surfaces.

Records of uptake and superheat temperatures taken during the passage will give a guide to the condition of the boiler, as deposits forming upon any heated surfaces of the boiler will reduce the rate of heat transfer; thus uptake gas temperatures tend to show an increase, while superheat temperatures decrease.

Where possible the boiler should be shut down at least 24 hours prior to cleaning, with if practicable the soot blowers being operated just before shut-down. When boiler pressure has fallen to about 400 kN/m^2, open blow down valves on drums and headers to remove sludge deposits. Finally empty the boiler by running down through suitable drains etc. Do not attempt to cool the boiler forcibly as this can lead to thermal shock. All fuel, feed and steam lines must be isolated, and the appropriate valves locked or lashed shut. Air vents must be left open to prevent a vacuum forming in the boiler as it cools down.

Internal Cleaning With boiler cooled, open the steam drum doors, followed by water drum doors and any bolted header plugs where fitted. Take care to avoid any remaining hot vapour or water, and allow the boiler to ventilate before making any attempt to enter. Light covers should be fitted in place of the manhole doors to protect the boiler interior when work is not in progress. With a man standing by outside the door, the boiler interior can be inspected to ascertain if cleaning is required.

Should cleaning prove to be necessary, remove any internal fittings required to provide access to tubes etc., keeping a record of any items removed. Also note that all attachment bolts are present, and that all are accounted for when refitting.

Where the boiler design permits, cleaning can be carried out by mechanical brushes with flexible drives; if these are not suitable, chemical cleaning must be used. After cleaning, flush the boiler through with distilled water.

Upon completion of cleaning, tubes etc. must be proved clear. Where access is available, search balls or flexible search wires can be used. Where neither is practical, high pressure water or air jets can be used, the rate of discharge from the outlet end being used to indicate whether any obstruction is present within the tube. Where necessary, welded nipples are removed to permit sighting through headers. With welded boilers the tubes must be carefully searched before welding takes place, and suitable precautions then taken to avoid the entry of any foreign matter into tubes etc. Where work is to be carried out in the drum, rubber or plastic mats can be used, with flexible wires attached and secured outside the drum so that they are not left inside when the boiler is closed up.

Check all orifices to boiler mountings to prove that they are clear, and ensure that all tools, cleaning materials etc. have been removed from the boiler. All internal fittings removed must be replaced. Fit new gaskets to all doors and headers, and close up the boiler.

All personnel working in the boiler must be impressed with the importance of the avoidance of any objects entering the tubes after the boiler has been searched, but that if a mishap should occur it must be reported before the boiler is finally closed up.

External Cleaning Spaces between tubes can become choked with deposits which are not removed by soot blowing. Where sufficiently loose they may be removed by dry cleaning using brushes or compressed air, but in most cases water washing will be necessary.

Washing will require hot water, preferably fresh, under pressure and delivered by suitable lances. The water serves two purposes, dissolving the soluble deposits and then breaking up and flushing away the loosened insoluble residue.

Once started, washing should be continuous and thorough, as any half-dissolved deposits remaining tend to harden off, baking on hard when the boiler is again fired, then to prove extremely difficult to remove during any subsequent cleaning operations.

Prior to cleaning, a bitumastic paint should be applied around tubes where they enter refractory material, in order to prevent water soaking in to cause external corrosion. Efficient drainage must be provided, with sometimes drains below the furnace floor requiring the removal of some furnace refactory. Where only a particular section is to be washed, hoppers can be rigged beneath the work area, and the water drained off through a convenient access door.

For stubborn deposits a wetting agent may be sprayed on prior to washing.

After washing, check that no damp deposits remain around tube ends, in crevices etc., removing any remaining traces found. In a similar manner remove any deposits in double casings around economiser headers etc., especially if they have become damp due to water entering during the washing process.

Ensure that all cleaning materials, tools, staging etc. have been removed, and any refractory removed has been replaced, after which the access doors can be replaced.

Run the fans at full power with air registers full open for some minutes to clear any loose deposits. Then dry the boiler out by flashing up in the normal manner. If this can not be done immediately, then hot air from steam air heaters or from portable units must be blown through to dry the external surfaces.

Q. Give a routine procedure for a soot blowing operation, suitable for use with high performance water tube boilers.

A. Before soot blowing is commenced, ensure that sufficient reserve feed water is available, that the soot blowing system is in good order with correct oil levels in blower gearboxes, and that where required bearings have been lubricated with a high temperature grease. Make sure that the boiler flame failure devices and low water level alarm and cut-outs are in full working order. Note the uptake temperatures if these are not already recorded. Then inform the bridge that soot blowing is about to start. They may require time to alter course etc. and will ring back when ready.

The soot blower main steam supply valve can now be cracked open and lines allowed to warm through and drain. The drain valves, if non-automatic, are then closed, and the steam supply valve is fully opened. In many cases two supply valves are fitted in series, with a drain between them to ensure that no steam can leak through to the soot blowing system when it is not in use.

Automatic combustion control can be shut off, or set to the blowing mode, whichever is applicable, and boiler operating conditions, such as superheat temperature, reduced if necessary. Shut off gas sampling lines to CO_2 recorders etc. and inert gas systems if fitted. Any gas or air control dampers should be set to their optimum position for soot blowing. The combusion air fans should be speeded up unless otherwise instructed.

115

Soot blowing can now begin, starting with the topmost blower and then, when this has completed blowing, proceeding to the lowest blower, operating this and the subsequent units in turn along the line of the gas flow. When the topmost blower is again reached, it is operated for a second time. Ensure that each unit performs correctly in sequence and over its full traverse. Operate any faulty units manually.

Throughout the operation maintain a constant watch on boiler operating conditions, and if necessary stop blowing.

When soot blowing is finished, inform the bridge. Then restore all the boiler functions to their normal operating conditions. Close the soot blower steam supply valves and fully open the drains. Check the reserve feed tank, and adjust the make-up feed supply as required to restore the water level in the tank. The gas sampling lines can be reopened, together with any inert gas system fitted. Note the uptake temperatures and compare them with previous readings; any marked variations can indicate some fault in the soot blowing system.

It is usual to operate soot blowers once every 24 hours, with rotary air heaters, where fitted, being blown every 12 hours.

Index

Accumulation of pressure, 83, 87
Acid attack, 63, 111
Air checks, 98, 102, 108
Air register, 36, 38, 92, 93, 94, 97, 99, 115
Air vents, 63, 69, 105, 106, 107, 114
Allborg boiler, 15
Alloy steel, 42, 51, 55, 57
Atomization, 61, 91, 94
Attemperators:
  air cooled, 41, 58
  water cooled, 41, 59
Automatic control, 17, 19, 38, 42, 94, 95, 96, 98, 103, 113, 115
Auxiliary stop valve, 67

Babcock Willcox boilers, 36, 38
Baffles, 13, 15, 24, 33, 35, 40, 52, 53, 57, 98
Balance connections, 60, 79, 81, 82
Bell mouth, 9, 13, 27, 45, 57, 112
Bend test, 6
Blow down effect, 83, 88, 89, 90
Blow down valves, 18, 30, 68, 105, 114
Blow off pressure, 21, 83, 85, 87, 89
Boiler cleaning, 114, 115
Boiler drum, 17, 28, 44
Boiler feed regular, 68
Boiler shell, 4
Bonded deposits, 44, 45, 56, 59

Carbon deposits, 92
Carry over, 112
Castable refractory, 102
Chemical cleaning, 114
Chemical dosing valve, 69

Chrome, 57
Chrome ore, 102
Circumferential joint, 8
Circumferential stress, 1, 2
Clarkson boiler, 12
Cochran boiler, 10
Collision chock, 7
Combustion:
  air, 91, 96, 98
  chamber, 4, 7, 11, 12, 18, 24, 73
  process, 91
Compensation, 3, 4, 107
Composite boiler, 22, 23
Condensation, 50, 74, 111
Conduction formula, 27
Consolidated safety valve, 89
Control dampers, 19, 38, 41, 53, 56, 60, 115
Controlled superheat boiler, 51
Corrosion, 9, 10, 28, 31, 44, 62, 63, 103, 111
Corten steel, 66
Cracking, 12, 107, 111, 112
Cyclone steam separators, 38

Dampers, 19, 38, 39, 41, 51, 53, 56, 57, 115
Density, 30, 33, 35
De-superheater, 53, 55, 60
Dew point temperature, 24, 25, 63, 64
Diaphragm water wall, 44, 46
Double casing, 33, 34, 36, 41, 59
Downcomers, 15, 28, 30, 32, 34, 36, 37, 39, 41, 42, 44
Drain valve, 63, 69, 99, 103, 115
D-type boiler, 35, 37, 54

117

Dual pressure boiler, 20

Easing gear, 73, 85, 86, 89
Economiser, 40, 52, 64
Enamel coating, 66
Erosion, 104, 112
ESD. I typeboiler, 39
ESD. II typeboiler, 41
ESD. III typeboiler, 42
Excess air, 42, 44, 47, 93
Exhaust gas boiler, 11, 15, 16, 21, 22, 25, 111

Fatigue cracking, 12
Feathering, 85, 87, 89
Feed water check valve, 67
Flame failure device, 12, 19, 101, 115
Flash off, 69
Forced circulation boiler, 16, 17, 18, 24, 25
Force draught fan, 18
Foster wheeler boiler, 35, 39, 40
Fuel oil system, 100
Full bore safety valve, 85
Full lift safety valve, 86
Furnace, 7, 10, 11, 12, 13, 14, 35, 43

Gamma ray equipment, 111
Gas/air heater, 8, 43, 64, 65, 108
Generating tube, 15, 24, 31, 33, 34, 35, 37, 38, 43, 44
Girder stay, 7, 8
Gusset stay, 11

Handhole door, 11
Header, 15, 27, 30, 33, 36, 39, 56
Header boiler, 33, 64, 109
Hemispherical furnace, 11, 18
Hollow column, 72
Hopkinson safety valve, 87
Hydraulic test, 30, 107, 109, 110

Igema gauge, 78
Ignition temperature, 92

Improved high lift, 83
Internal cleaning, 106, 114
Integral furnace boiler, 36, 84
Internal inspection, 106, 108, 110

Joint efficiency, 2

Kaldo steel, 5
Kinetic energy 96, 102

Latent heat, 19, 49, 50
LD steel, 5
Longitudinal joint, 1, 2, 8, 29
Louvre plate, 77
Low water level alarm, 19, 68, 80, 81, 83, 115

Main stop valve, 67
Manhole door, 2, 4, 8, 11, 13, 15, 18, 20, 29, 30, 38, 40, 105, 106, 108, 109
Manoeuvring, 51, 55
Membrane water wall, 46
Mica, 76, 77
Multi-loop economiser, 62
Multi-loop superheater, 39, 41, 45, 57
Multi-tubular boiler, 7
Molybdenum, 32, 57
Monolithic refractory, 102
Mono water wall, 41, 43, 46

Natural circulation,
Nozzle reaction, 86

Ogee ring, 10, 11, 12, 13
Oil contamination, 19, 20, 112
Orifice plate, 61, 94, 104
Overheating, 10, 11, 12, 31, 53, 54, 55, 58, 73, 111, 112

Package boiler, 18
Parallel flow, 17, 42, 52, 64
pH value, 111
Plate glass gauge, 75, 76
Pop action, 89

Pressure jet burner, 93
Primary combustion air, 92, 97
Primary flame, 92
Primary superheater, 40, 41, 44, 59
Priming, 112

Quarl, 11, 92, 99
Quick-closing valve, 100, 113

Radiant heat, 11, 15, 31
Radiant heat boiler, 31, 32, 44, 65
Raising steam, 105, 107
Recirculating valve, 100
Reflex water gauge, 75
Refraction, 71, 73, 75
Refractory, 11, 12, 15, 36, 38, 40, 41, 46,
    48, 99, 101, 102
Relief valve, 63
Remote reading gauge, 78, 108
Riser, 32, 36, 44
Riveted joint, 3, 5, 6, 8, 11, 14, 28, 107,
    112
Roof firing, 42, 112, 113
Rotary air heater, 44, 65
Rotary cup burner, 95

Safety valve, 19, 21, 55, 67, 73, 83, 85,
    86, 109, 111
Salinometer valve, 69, 105
Scale, 8, 14, 20, 28, 106, 112
Scotch boiler, 7, 8, 18, 27, 64, 70, 105,
    106 107
Screen tube, 31, 33, 35, 37, 38, 41, 52,
    57, 111
Scum valve, 69
Seal welding, 10
Secondary combustion air, 93, 97
Secondary flame, 92
Secondary superheater, 40, 41, 42, 44, 59
Selectable superheat boiler, 38, 55
Settling tank, 99
Single boiler casing, 41, 44, 47
Smoke box, 8, 11, 14, 18
Smoke tube, 8, 9, 14

Sodium, 52, 63, 64
Soot blower, 41, 43, 63, 64, 65, 66, 102,
    108, 115
Soot blower master valve, 70, 103, 115
Spanner boiler, 14
Spheroid boiler, 11
Spheroidal furnace, 11
Spray type attemperator, 45
Spray type de-superheater, 51, 61
Steam/air heater, 66
Steam blast jet burner, 96, 99
Steam drum, 15, 16, 22, 28, 30, 33, 35,
    38, 39, 41, 43, 59, 60, 67, 68, 78, 81,
    108, 111
Steam/steam generator, 19
Steaming economiser, 45, 64
Stay, 5, 7, 11, 13, 15, 18
Stay tube, 8, 9, 11, 12, 15, 18
Stub tube, 29
Studded tube, 36, 38, 46
Sulphur, 63, 93
Superheater circulation, 52, 53, 55, 58,
    60, 67, 69
Superheater circulating valve, 69
Superheater support tube, 32, 45, 51, 52,
    54, 56, 57, 111
Superheater temperature control, 33, 34,
    36 38, 40, 41, 42, 44, 45, 51, 53, 55, 58
Superheater tube, 32, 39, 52, 55, 56, 58
Survey procedure, 110
Suspended flame, 91
Swirl plate, 94
Swirlyflo tube, 15

Tangential firing, 43
Tangential water wall, 44, 46
Tank boiler, 7, 10, 14, 23, 25
Tensile test, 6
Test piece, 6, 29
Testing of safety valve, 110, 111
Testing of WL gauges, 71, 74, 79, 105,
    108, 110, 111
Thermal efficiency, 22, 49
Thermodisc valve seat, 89

Thimble tube, 12, 13
Thin shell formula, 27, 29
Tip plate, 92, 99
Tube attachment, 9, 10, 11, 12, 13, 19, 27, 29, 32, 36, 44, 45, 54, 56, 57, 62, 64, 65
Tube plate
Tubular air heater, 64, 65
Tubular WL gauge, 70, 71, 73
Turn down ratio, 94, 95, 96
Two-drum boiler, 34, 38

Ultra-sonic test, 110
Uptake, 8, 13, 15, 18, 23, 35, 44, 57, 62, 63, 64, 116
Uptake fire, 63, 64, 65
Uptake fouling, 63, 64, 65, 66, 99
Uptake heat exchanger, 63, 64, 65
Underfloor tube, 35, 36, 37, 39, 41

Valve lift, 83, 85, 87
Vanadium, 52, 102

Venturi, 97
Vertical boiler, 10, 12, 13, 14, 16, 23
Viscosity, 91, 99, 100
Vortices, 91, 92, 99

Waste heat boiler, 17, 22, 23
Waste steam pipe, 83, 85
Water circulation, 8, 13, 15, 16, 18, 20, 24, 25
Water drum 30, 35, 37, 38, 39, 41, 43
Water level gauge, 68, 70, 71, 73, 75, 76, 78
Water wall, 15, 31, 34, 35, 37, 38, 39, 41, 44, 46
Water washing, 115
Welded joints, 8, 10, 11, 13, 14, 15, 19, 28, 38, 45, 46
Wrapper plate, 28, 29

X-ray examination, 29

Y-jet burner, 96